March Farm

Season by Season
On a Connecticut Family Farm

by

Nancy McMillan

Photographs by Stuart Rabinowitz
and Jack Huber

For Sandi,
With all best wishes —
Nancy McMillan
December 2016

March Farm: Season by Season on a Connecticut Family Farm
©2012 by Nancy L. McMillan
Printed in the U.S.A.

McMillan, Nancy.
    March Farm: season by season on a Connecticut family farm / Nancy McMillan ;
photographs by Stuart Rabinowitz and Jack Huber.
        p. cm.
    Includes bibliographical references.
    ISBN: 978-0-9884299-2-5
    1. Farms, Small—Connecticut. 2. Sustainable agriculture—Connecticut.
3. Farmers—United States. 4. Connecticut—History, Local.   I. Title.
S451.C7 M63 2013
635—dc23

ISBN 978-0-9884299-2-5

Book design by Huber Design Group

Whistling Hawk Press
P.O. Box 66
Bethlehem, CT 06751
www.marchfarmthebook.com

Second Printing

Dedicated to the farmers at March Farm, past and present

TABLE OF CONTENTS

Opposite: View of former dairy barn, converted into living quarters by Tom March, Jr.

The Present

From my bedroom window at four o'clock on a calm summer morning, I can hear the clunky murmur of a tractor half a mile away. I recognize the sound. The farmers at March Farm are up and working.

Munger Lane winds through and around March Farm. I drive this road several times weekly, along the shaded straightaway that marks the eastern boundary. From here, under the row of tall arching maples, the orchards slope down the hill, then up the next, in varying shades of green. The view transports me out of my daily life.

Around the bend and down the hill sits the farm store, a low building sided barn-style with vertical planks stained brown. Brightly painted wooden signs of apples and blueberries and peaches decorate its front. Flower boxes filled with zinnias and marigolds in carnival colors stand on either side of the entrance. Hanging pots of nasturtiums brighten the narrow porch at the end of the building.

Inside is the cornucopia of midsummer: indigo blueberries in green cardboard cartons; plump peaches, giving slightly to thumb pressure; red globes of tomatoes piled next to bunches of fresh basil, smelling of salt and earth.

Driving by the farm releases some tension inside me, allowing me to inhale a little deeper. Driving by any farm anywhere makes me feel better. How does it inspire such comfort? What is so reassuring about the presence of the farm landscape? Is it because it signals to our primal hunter-gatherer that our food supply is safe, an issue that is becoming more critical today? Do the open fields surrounded by woods speak to some deep genetic need in us?

That is the theory of biophilia, put forth by E. O. Wilson, a Harvard entomologist. Wilson defines biophilia as "our love of living things, our innate affinity with nature and tendency to focus on life and lifelike processes." A related field, ecopsychology, seeks to develop a "mature ecological ego which weaves a sense of ethical responsibility to the planet into the fabric of social relations and political decisions." Perhaps without realizing it, family farmers like the Marches practice a form of ecopsychology, stewarding the earth to produce food for the community at large. This family has survived for nearly one hundred years doing just that.

March Farm is a 150-acre fourth-generation family farm in northwest Connecticut. This book describes a year in the life of March Farm. Accompanying photos show the beauty and practical realities of farm life. Recipes using the farm's crops are also included, as well as informational sidebars.

I wrote this book because I fell in love with the farms in Bethlehem and became passionate about saving them. The loss of farmland is a problem in this state, as it is across the country. If we are to preserve our heritage and keep our food source safe, as well as support our communities, we need to both understand and cherish our farms.

Opposite: Farmworkers picking corn at Swendsen Farm in Bethlehem, where the Marches lease acreage.

Situated in southern Litchfield County, Bethlehem is a gateway to the Litchfield Hills and northwest Connecticut, an area known for its traditional New England scenery: small towns boasting village greens and historic homes surrounded by rolling countryside of woods and fields. The landscape of Bethlehem is still fifty percent forest and farmland, lending it a rural beauty that never ceases to elicit comments from visitors.

I know exactly what they mean. Every time I drive home from points south, my heart lifts at the sight of three hay fields, like golden pillows resting side by side in the summer light. They sit at the top of the long incline on Route 132, just before the ninety-degree bend that brings into view the Parmelee farm spread out below and beyond it, the blue hills filling the western horizon.

But I also know how many farms have been lost here, succumbing to the pressure of development and the economic realities of farming. In the 1950s, at least two dozen active dairy farms filled the town. They used to say there were more cows than people in Bethlehem. Over the years, dairy farmers have been squeezed out of business due to deteriorating market conditions. Now only a handful remain. The farms that are making it, like March Farm, have diversified.

The disappearance of farms is not confined to this corner of the state. Every year Connecticut loses between 7,000 and 9,000 acres of farmland. If this rate continues, there will be no farmland left to save by 2040. In 1978 the state's Farmland Preservation Program, administered by the Connecticut Department of Agriculture (CTDOA), was established with a goal of preserving 130,000 acres of the state's farmland. Thirty-four years later, only 38,078 acres have been preserved under this program, largely because the annual appropriation of funds in the state legislature has not matched the intent of the bill. Although 291 active farms have been preserved through the program, there is still a long waiting list.

The CTDOA preserves farmland by acquiring development rights to agricultural properties, giving farmers a realistic alternative to selling their farm for residential development. Participation in the program is voluntary. The farms remain in private ownership and continue to pay local property taxes. Farmers can reinvest proceeds from the sale of development rights back into the farm; they can also sell or lease the farm. A permanent restriction on nonagricultural uses is placed on these properties, no matter who owns or uses the land.

Loss of farmland is a nationwide epidemic. Every year since 1992, the United States has developed more than one million acres of farmland, which equates to roughly 3,000 acres per day. That is 1,562 square miles of land, a bit larger than the size of Rhode Island, lost yearly, never to be reclaimed.

The financial viability of smaller farms is also of concern. While small family farms like March Farm account for most of the nation's farmland and farm assets, large farms and nonfamily farms produce the largest share of agricultural output. As of 2004, small farms make up ninety percent of the farm assets while producing only twenty-five percent of farm production. Large farms account for fewer than eight percent of active farms while generating sixty percent of the production.

There is good news, though. People are now paying closer attention to how their food is produced and are more concerned about saving the farms that are left. There is a growing awareness that, as the bumper sticker on my car reads, "No Farms, No Food." In magazines, on the radio, in newspapers, and especially on the Internet, food and farms are topics of regular discussion. The farm bill is usually passed every five years or so without much attention from the average citizen. The latest bill, formally called the Food, Conservation, and Energy Act of 2008, garnered intense focus from activist groups across the country. When this usually ho-hum legislation becomes worthy of a *New York Times* op-ed by Michael Pollan, awareness is changing. A new movement has arisen focused on the connection between food production and the health of consumers, communities, and the environment.

This movement, like good farming, is diversified. It has found spokespeople in Michael Pollan *(The Omnivore's Dilemma, In Defense of Food)*, Alice Waters *(The Art of Simple Food)*, Barbara Kingsolver *(Animal, Vegetable, Miracle)*, Bill McKibben *(Deep Economy)*, and Paul Roberts *(The End of Food)*, and continues to draw upon the wisdom of Kentucky farmer and philosopher Wendell Berry, who wrote, "Eating is an agricultural act." This quote has become the starting and ending point of the national conversation on food and farming.

Local food has become desirable food. Approximately twenty-five percent of today's consumers seek out food locally grown by family farmers like the Marches. The growth of farmers markets nationally is a testament to this: a 114 percent increase from 2000 to 2010. In Connecticut, home to over 100 farmers markets, the number increased by 50 percent from 2005 to 2008, with more than half of that increase occurring between 2007 and 2008. Community-supported agriculture (CSA) farms, in which consumers purchase shares for the season and in turn receive a weekly harvest of produce, have waiting lists. Some here in Connecticut have even closed their waiting lists.

Programs linking local food and institutions are springing up. Plow to Plate is a regional grassroots initiative supported by New Milford Hospital, which is setting an example by committing to changing its in-house food service sourcing to local sustainable foods. Since 2003 the Yale Sustainable Food Project has offered an organic menu sourced locally whenever possible at the cafeterias on campus. The project was initiated with the help of chef Alice Waters and has spread, through student envy, to the eleven cafeterias on campus as well as to other schools in the state, such as the University of Connecticut, Trinity College in Hartford, and The Unquowa School in Fairfield. People and institutions are looking for ways to connect to their daily sustenance.

In *Deep Economy*, Bill McKibben writes, "A tomato from the small farmer at the end of your suburban road takes less fuel to transport, and a tomato from the farmer at the end of your suburban road tastes better. But it's more than that — it's better because it comes from a … farmer down at the end of your suburban road. Getting that tomato — from his farm stand, from a farmer's market, from your CSA share, even from a bin at an enlightened supermarket — requires you to live with a stronger sense of community in mind."

Michael Pollan reminds us of the layers of meaning inherent in what we eat in *In Defense of Food*: "We forget that, historically, people have eaten for a great many reasons other than biological necessity. Food is also about pleasure, about our community, about family and spirituality, about our relationship to the natural world, and about expressing our identity. As long as humans have been taking meals together, eating has been as much about culture as it has about biology."

The authors of *The Cluetrain Manifesto: The End of Business as Usual* contend that twenty-first-century consumers who grew up using the Internet are accustomed to having a conversation about everything and are interested in having relationships as part of their buying experience. So when they participate in a CSA, purchase from a farmers market, or frequent a pick-your-own (PYO) farm, they are building a relationship as well as purchasing a food product.

March Farm is benefiting from and responding to this trend. The Marches currently participate in the Litchfield Hills Farm-Fresh Market held on Saturday mornings for most of the year, and were involved in the formation of the market. More PYO crops such as strawberries and cherries have been added to their orchards; their farm store has been expanded. Over the past several years, from July through October, the crowds at the farm have grown markedly.

The March family has developed the agritourism element of the farm by making it a destination for families. A hay bale playground with picnic tables under a pavilion, a small pond-side barnyard, a corn maze, farm hike, and hayrides all are recent additions to the farm. An autumn apple festival provides additional agricultural experiences. The steady stream of customers comes for more than the blueberries, peaches, apples, corn, and tomatoes. They come to experience the peace and beauty of the farm and to interact with the farmers.

In the process of writing this book, I have become more conscious of where my food comes from and the costs associated with growing it: the cost to me, to the farmer, to the environment. The pleasure of using fresh local ingredients in my kitchen matters to me now. My appreciation of the seasonal cycle of local foods has deepened. The fragrant juice of a summer peach sheets my paring knife; a blueberry right off the bush bursts with flavor; an autumn apple from the tree delivers a crisp crunch; the picked-that-morning corn in late summer is sweet and tender. Connecting with the food March Farm produces gives me roots in my community, and roots in the seasonal cycles of my own life.

My hope is that this book will move the reader to pay more attention to the journey of his or her daily nourishment from seed to table. Reconnecting with this basic human need can reap surprising pleasures and add joy and meaning to everyday life. Cherishing and preserving our farms is important to all of us, as communities, states, regions, and a nation. Small family farms are iconic symbols of values held deep in our cultural psyche, as well as protectors of our food supply. They need us for their survival, as we need them for our own.

Opposite: Siblings getting ready to fill their baskets with blueberries.

CHAPTER ONE

Autumn: Harvest

In the Apple House

"Do you mind a little music?" Tom March asks me on a cool, clear morning in late September. He starts up an old record player, the kind with the latched lid your gym teacher used on square dance days. A scratchy version of Ray Conniff's voice emerges from speakers mounted on the wall above his record collection. We listen to "The Look of Love" while Tom works.

It is six a.m. and we are in the apple room, home of the U-shaped conveyor belt that aids the apple-sorting process. Early light filters through plastic-covered windows. Tom tells me the conveyor belt is from the thirties, and it looks it, sturdy but old-fashioned in design and style. "Older than me," he says.

He walks outside and hops on a forklift, steers its flat prongs under a large wooden bin of McIntosh apples, and drives through the wide door of the apple house. The bin is lowered directly into a square tank, the 1,500-gallon water dumper. Each 600 pound bin of apples takes about an hour to process.

The apples bob up as the bin sinks into the tank. They float to the surface, turning the water into a moving wave of red spheres. They are picked up on an inclined roller beneath the water's surface that carries them to the conveyor belt. After the first bend, the apples disappear into the enclosed length of the polisher. Whirling black brushes make a small whirring sound. Shiny apples emerge and bump through the next bend to the sorter.

Above: Apple bins and baskets on the second-floor loading dock of the apple house. Opposite: View of the farm from the western apple orchards.

Initial size sorting is done by spinning rubber cones along the edge of the belt which bounce out certain size apples into an area bordered by a diagonal metal divider. The apples are then divided into three categories: firsts must be immaculate; seconds may have some small marks; thirds are misshapen with too many blemishes for human use. They'll be sold as "horse apples" to the many horse owners in the area.

Tom maintains a strict standard for his firsts. They must be picture-perfect, and these Macs qualify, vivid crimson orbs with patches of green near the stem. The seconds are good enough for my needs — fine for eating and excellent for baking. Thirds are packed in small wooden crates. A faint sweet smell fills the air. Although Tom sorts rapidly, his bear paw hands grabbing two apples at a time, there is an easygoing pace to the work. He packs the firsts into baskets in a consistent circular pattern that displays the apples nicely. This year the apples are so large they top the basket generously, and customers receive a good value for their money, as the apples are sold by basket size, not weight.

The season runs from late August to November. In 2007 Tom hand-sorted 60,000 pounds of apples. These September Macintoshes will appear in the store among the varieties now in season: Idareds, Cortlands, and my eating favorite, Honeycrisp. Developed in 1960, the Honeycrisp is a relatively new apple, compared with the Baldwin, which has been around since the mid 1700s, or the Northern Spy, discovered in 1800. The Honeycrisp is a cross between a Macoun and a Honeygold. With the crispness of a Macoun and a coarse, creamy-colored flesh, the Honeycrisp is one of the juiciest apples I've ever eaten.

Tom March grading apples.

Apples are the fruit with the largest range of varieties, colors, tastes, and textures. From the lime green tartness of the Granny Smith to the crispness of the red orange Gala to the deep wine color and baked sweetness of the Rome Beauty, there is an apple to satisfy every taste and need. Sweet, tart, juicy, firm. Eating, baking, storing, cider making. Twenty-four kinds of apples were recorded in ancient Rome; locally, Native Americans mentioned apple orchards in a 1733 deed. In 1872, a thousand varieties growing in orchards in this country were listed by American pomologist Charles Downing. Today there are 7,500 known varieties worldwide and 2,500 grown in the United States.

Yet even with a mere fourteen to choose from at March Farm, people tend to gravitate toward their favorite. "People are funny about their apples," says Lynn Horvath, working the counter in the farm store. "They come in here and ask me what the best eating apple is. Well, it all depends on what you like. They get stuck on their McIntoshes. Try a Rome, I say — be adventurous. There are better cooking apples than Macs." She consults the chart behind the register that categorizes apples by their best use. "You see, Macs have a question mark here, under baking. Romes are much better."

The fruit's symbolism is as rich as its variety. Apples appear in mythology, religion, and legends. Think of Eve and the forbidden fruit, Aphrodite and the golden apple that eventually brought down Troy; William Tell's accurate shot, Snow White's bite into the poisonous fruit, Isaac Newton's discovery under an apple tree. For over two thousand years apples have served as symbols of beauty, passion, temptation, love, fertility, and goodness.

And good health. According to legend, King Arthur ate apples to achieve immortality; the Greeks wrote of apples on the tree of life in the garden of Hesperides. And who hasn't heard the advice "An apple a day keeps the doctor away." Apples contain vitamin C, antioxidants, and pectin. All this in a compact package sturdy enough to hold up in a lunch box.

Apple Crisp

Preheat oven to 350 degrees.

Filling:

- 8 cups (2 lbs.) apples, any variety, chopped (peeled if necessary)
- 2 T sugar
- 1 T lemon juice
- 1 tsp. cinnamon

Toss together in large bowl. Pour into three-quart glass or ceramic baking dish. Set aside.

Topping:

- 1 cup flour
- ½ cup old-fashioned rolled oats
- ¼ cup sugar
- ¼ cup brown sugar
- 1 tsp. cinnamon
- ½ tsp. ground cloves
- ½ tsp. ground ginger
- ½ cup unsalted butter, melted
- ½ cup pecans or walnuts, chopped

Stir together flour, oats, sugars, cinnamon, cloves, and ginger. Stir in butter until evenly moistened crumbs form. Stir in nuts. Spoon over filling. Place on middle rack in oven. Bake until juices are bubbling and top is lightly browned, 30 to 40 minutes. Serve with a scoop of vanilla ice cream or a dollop of homemade whipped cream.

At the farm, apples are the third-largest crop, after tomatoes and corn, and the oldest. The first orchards were planted in the fifties, and pick-your-own began in the nineties. Now, every sunny weekend from September to November, the newly expanded parking lot next to the store is jammed with cars. Ben March, Tom's son, who has returned to work the farm, has added new signs showing a map of the orchards above the store, and road signs directing pickers to the orchards up on the hill. For those seeking convenience, ready-to-buy baskets and bushel bags line the farm stand shelves. Cider is for sale, along with cider doughnuts, if you get there early enough. Homemade mulling spices are packaged by teenagers working behind the counter during rare lulls. For most of the day, the lines are long in front of two registers.

And you'll see Tom in the stand, bringing in apples from the apple house across the street, talking to customers as he replenishes bins or sorts through apples. He appears at ease as he works and converses with people in the store, but his hands, like Tom himself, never stop moving. For there are always more apples to sort.

Farm Field Trip

On a brisk October morning, two busloads of fourth graders arrive from Carrington Elementary School in Waterbury. The children spill out of the bus and gather on the hill above the farm store, just below the apple orchards. They are visiting the farm as part of a school field trip. The air is warm enough for short sleeves. The sky is cloudless, a solid blue umbrella. In the distance crows, like flapping black flags, pursue a hawk, cawing insistently until the hawk escapes in a high gliding flight.

Waterbury, once known as the Brass City because of its long history as the center of the nation's brass industry, is the closest city to Bethlehem, eleven miles southeast of March Farm. With a population of 110,000, it is the fifth-largest city in Connecticut. Its population is a diverse mix of ethnic, cultural, and religious backgrounds. This school group reflects that rich diversity.

I join one of the groups and stand at the back. A chubby boy in a striped T-shirt stands in front of me as we listen to Sue March, Tom's wife, talk about the apple trees. Ben then steps in to describe the morning's activities.

"And the last thing we'll be doing is picking apples," he tells the group.

All the children respond the same way: "Oooooh."

The boy in front of me points at a large bird circling above us, the black underside of its wings banded with silver. The boy's eyes disappear into a squint in his brown face.

"Look at that," he says to no one in particular.

"It's a turkey vulture," I say. Vultures, with their impressive six-foot wingspan, are a common sight in rural areas.

He stares at it a moment longer, then snakes through the crowd to find his friend. They both look at the bird. I wonder if they've ever seen a vulture before.

The children tour the apple cooler behind the farm store and the apple-sorting room across the street, where the apples are washed, polished, and sorted. They lean in over the shiny apples and watch Tom sort the fruit by hand. We head for the two-story building behind the apple room where cider is made. The children sit down on the floor and look up at Ben as he explains the cider operation.

Making cider is a complicated process. A forklift dumps a bin of apples onto a table on the left-hand side of the large room; then an angled conveyor belt delivers them into the top of a large box. The apples are dropped into a high-speed grinder, which mashes them into an applesauce mix. Every part of the fruit is used. The mash is sprayed onto square canvas screens and then covered with a plastic form. The screens are stacked high and squeezed by a 1,200-pound press, releasing the juice into the 500-gallon cider tank.

One 600-pound bin of apples yields fifty gallons of cider. Ten bins fill the tank with cider. To meet health codes, a small amount of the preservative sodium bisulfate is added. The cider is then bottled in plastic jugs with color-coded caps for dating and refrigerated until sold. Leftover apple mash becomes applesauce for animals and livestock at nearby farms. The children sip cider, then gather on the lawn in front of the brick red farmhouse across the street from the farm store.

Opposite: View from the peach orchard in the northwest corner of the farm.

Ben March (rear) guides children from Carrington Elementary school in Waterbury on a tour of the apple house.

One of the children asks Ben, "What do you like most about doing this?"

"Being outside," he says.

The students look for walnuts on the ground, fallen from two large trees framing the farmhouse's front door. Bags are passed out for apple picking, and the children scatter into the orchard like a flock of birds set free.

Ben March talks about the success of the program. "We are raising awareness that family farms still exist in Connecticut. To make it an educational experience, we like to show children the entire process of a typical harvest. We hope that they'll not only have a positive experience here, but that they'll come back on the weekends with their parents."

"Even driving up here from Waterbury to Bethlehem and seeing the countryside was a unique experience for many of them," says their teacher. I think about lives bound by asphalt and concrete buildings with views of more of the same, bird sightings limited to rooftop pigeons or seagulls circling city restaurants. How often do these children experience nature up close?

My childhood experience of nature in southern Connecticut was formative. I remember the woods behind our suburban house. It was a thrill to pick my way down the steep hill, past the stockade fence that marked our boundary, all the way to the bottom, to the Rippowam River. In the summer my friends and I swam, watching for rocks, aware of some form of danger. We reveled in the cold water and the freedom from adults, venturing past the boundaries of everyday life and closer to nature. I spent time alone there, too, sitting on a large boulder next to the moving water, thinking. This was a safe oasis.

When we took trips to western Pennsylvania, where my grandparents lived, the rolling hills revealed farm after farm, like huge garden plots as far as I could see. For this view I would rouse myself after an eight-hour drive and roll down the window to take it all in.

I'm not alone in this response. In a survey of cultures around the globe, Stephen Kellert, a leading expert in biophilic design, along with other researchers, found that people were attracted to the same kind of landscape: open grassland with some trees and woods along the edge. The theory is that this landscape kept humans safe and helped us find food and, as a result, we feel better, more secure and calmer, in such conditions.

Pulitzer Prize–winning scientist E. O. Wilson coined the term *biophilia* to name this deep, instinctual affiliation. In his essay "Biophilia and the Conservation Ethic," he writes, "For more than ninety-nine percent of human history people have lived in hunter-gatherer bands totally and intimately involved with other organisms … In short, the brain evolved in a biocentric world, not a machine-regulated one." He suggests we are genetically coded to interact with the natural world. Researchers are beginning to collect scientific data that supports the current anecdotal data demonstrating that children with attention deficit disorders become calmer when spending time immersed in nature.

The Waterbury schoolchildren, having filled their apple bags, sit on the lawn in front, munching their harvest. I look for the boy who spotted the turkey vulture. I wonder if he'll remember what it feels like to see the large bird, wings extended, floating above him. I wonder if being on the farm stirs on the edge of his consciousness some ancient response buried deep in his DNA.

Perhaps the experience of nature is just as important as the experience of farm life for these children. "Nature presents the young with something so much greater than they are," Richard Louv writes in *Last Child in the Woods*. "Immersion in the natural environment cuts to the chase, exposes the young directly and immediately to the very elements from which humans evolved: earth, water, air, and their living kin, large and small."

Louv believes an early connection to our natural environment serves children and the earth. "If children do not attach to the land, they will not reap the psychological and spiritual benefits they can glean from nature, nor will they feel long-term commitment to the environment, to the place. This lack of attachment will exacerbate the very conditions that created the sense of disengagement in the first place — fueling a tragic spiral, in which our children and the natural world are increasingly detached." Louv has coined the term *nature-deficit disorder* to describe this condition.

Maybe for these city children this taste of farm life will provide a tiny memory that will lodge in their brains and emerge later, prompting them to seek a connection to nature. It could begin with an awareness of how the apples piled up in the produce section of their neighborhood supermarket are grown on a tree in an orchard only a half-hour bus ride away. Perhaps that recognition will be an entry point into an expanded awareness of the natural world and how their individual life fits into it.

A March Farm pie fresh out of the oven.

Baking Pies

The pie itself is a beauty. Cinnamon-scented apples sit under a slit crust bordered by a thick, neatly fluted edge. The first forkful delivers tender but toothsome fruit wrapped in a buttery crust. The fruit is just sweet enough to contrast with the slight saltiness of the pastry. The crust has a lingering presence and an excellent crumb.

A March Farm apple pie is one we often bring when traveling to visit distant friends. Even sophisticated city dwellers say the pie is "the best we've ever tasted." The empty pie tin the next morning is proof.

I bring a latecomer's enthusiasm to pie eating. I didn't learn to appreciate its pleasures until I was an adult. My mother was a wonderful woman in many ways, but cooking was not her strong point. The Thanksgiving pies that came out of her kitchen were adequate to satisfy a sweet tooth but not enough to develop a taste for pies. It was only when I started to bake them myself that I came to understand the complexity of the flavors and textures pies offer.

I arrive at the commercial kitchen in the basement of the farmhouse one sunny, crisp November morning when the bakers are at work on the big push for Thanksgiving orders. The kitchen is full of smells: sugar, cinnamon, butter, baking apples. Sue March is assisted by Lynn Horvath, from Bethlehem, and Becky Jimmo, from nearby Woodbury. The three women wear long green or red aprons tied around their waists and are elbow deep in slicing apples and assembling pies. They work at a stainless steel table that runs down the center of the kitchen. As Lynn stirs the cinnamon sugar into a large bowl of apples, her spoon makes a tiny ping against the stainless steel bowl. Flawless pastry is draped over a pie plate, then heaped with apples, and the crust is carefully laid on top.

Lynn, a tall, slim, dark-haired woman, presses her finger along the rim of a pie pan as she flutes the edge. "Every flute is different, like a fingerprint. I try for the copycat crust, like Sue's. I think it looks like a sunflower," she says. "Baking is my therapy." She looks up and smiles. "I guess I should be cured by now." All three women laugh.

Surprisingly, apple, not pumpkin, is the most popular pie at Thanksgiving. Although the orchards of March Farm produce fourteen different kinds of apples, Sue prefers, Rome for baking. "It gives a mellow flavor. I like Cortland, too, and I'll use anything we have, but Rome is my favorite." This year orders are the highest ever, at 320, beating last year's 299, reflecting the increased business at the farm over the past few years.

The women are relaxed, and the pace, like other work on the farm, feels leisurely yet productive. Pies are prepared five at a clip until one of the two ovens can be filled. One large convection oven fits twenty pies; the second, smaller one holds eight. Once in the oven, the pies are spun around so they bake evenly.

Just watching these women makes me want to don an apron and join them in the ritual of preparing Thanksgiving pies. The women open the oven doors wearing long thick mitts up to the elbow and pull out pies browned to perfection, crusts stained with bubbling juices. A heavenly smell fills the kitchen, and I am filled with a yearning for the comfort of family and food at a Thanksgiving table.

The March Farm pies, known for their consistent deliciousness, represent the care that goes into all the food produced here. Their reputation, and the future of the farm, rides on the quality that keeps customers coming back. Once you taste these pies, you'll be back for more. They are the next best thing to homemade.

From left: Bakers Sue March, Lynn Horvath, and Becky Jimmo.

Cider Doughnuts

The Sanitary Donut Machine holds ten pounds of doughnut mix along with ten and a half cups of cider. The crank is loaded on the mixer, then swung over the pan of vegetable shortening, already heated to 375 degrees.

The doughnuts making a plopping sound as they drop into the oil, which bubbles up briefly with each doughnut. Sue shakes the screen in the fryer to make them all float, then flips them in a minute or so. They cook up quickly. Sue's experienced eye knows by their color when they're done.

The screen is lifted out and the doughnuts drained of oil. Blue shop towels are
laid on the large draining sheet to keep
them from getting too greasy. Sue scrapes
the mixing bowl with a rubber scraper,
using the remainder of the batter to make
doughnut holes. The operation is tidy and
efficient. "I like to keep it neat because I
hate sticky, sloppy stuff," Sue says.

Even before the sugaring operation,
these doughnuts beg to be eaten.
With their crispy-crunchy outside,
and warm, dense, cakey middle, they
are doughnut perfection. Sugaring
the doughnuts is the last step. Lynn
uses a large circular plastic container
to mix up the cinnamon sugar. The
doughnuts are loaded in and the
cover sealed. Lynn lifts the container
and spins it slowly, to avoid a "blizzard of sugar."

March Farm Blueberry Tea Cake

This recipe is often requested by Sue's family.

- 2 cups flour
- 2 tsp. baking powder
- ½ tsp. salt
- ½ cup butter, softened
- ¾ cup sugar
- 1 egg, unbeaten
- ½ cup milk
- ½ tsp. vanilla
- 2 cups blueberries, fresh or frozen

Sift together flour, baking powder, and salt. In a separate bowl, cream butter, gradually adding sugar. Add egg, milk and vanilla; beat until smooth. Add dry ingredients; fold in berries. Spread batter in greased square pan (8 or 9 inches). Sprinkle with crumb topping. Bake at 350 degrees for 40 to 50 minutes.

Crumb Topping:
- ½ cup sugar
- ½ cup flour
- 1 tsp. cinnamon
- ¼ cup butter

Mix sugar, flour, and cinnamon. Cut in butter to form coarse crumbs.

No Farms, No Food

At breakfast I count how many items on my plate would be approved by a "locavore," an eater who endeavors to eat only locally grown foods. My oatmeal? No, but at least it's organic. The maple syrup? Yes, bought from a farmer on a Vermont vacation. Toast? Yes, from the fabulous Bantam Bread Company in the next town. The milk? Also yes, from Stony Wall Dairy Farm in northern Connecticut, which delivers weekly to March Farm. Butter? No, but not bad so far.

The applesauce? Yes! Cooked on my stove with apples from March Farm. Locally grown and locally cooked. I feel more satisfied and appreciative as I eat. And yes, I admit, a bit virtuous, too.

Every year Oxford University Press chooses a Word of the Year. In 2007, it was *locavore*. "The word shows how food-lovers can enjoy what they eat while still appreciating the impact they have on the environment," said Ben Zimmer, editor for American dictionaries at the press. "It's significant in that it brings together eating and ecology in a new way."

Being a locavore means different things to different people. A strict adherent might keep a one hundred-mile radius, which would mean that not only is her bread baked within that radius, but also that all ingredients from flour to salt were produced in that same area. Others use a broader definition, limiting only the final product to a one hundred-mile radius. Some aim for a portion of their food to be local. The Park Slope Food Coop, in Brooklyn, is the largest member-owned and -operated food co-op in the country. As of 2010 it was able to purchase 134 produce items within a 150-mile radius. Some locavores allow leeway for imported goods such as olive oil, coffee, and chocolate.

It's not the exact definition that matters, but the impetus behind the philosophy: transported food isn't as fresh and it's not allowed to ripen before harvest, and the transport and extra processing add carbon emissions to the atmosphere, contributing to pollution and climate change. Another reason to venture into locavorism is the desire to support local farmers and local economies and to realign our bodies to the natural cycle of seasonal produce. A recent and compelling reason is one that's come under public scrutiny: food safety.

In 2006 an outbreak of *E. coli* in spinach was traced to Natural Selection Foods LLC, of San Juan Bautista, California. The company is, according to its website, "North America's leading supplier of specialty salads." It is the parent company of Earthbound Farm which produces organic greens, the kind you see in the purple and green plastic bags and boxes in grocery stores. The company also buys, processes and packages conventionally grown greens for Dole, Trader Joe's and Sysco, among others.

This marked the ninth time since 1995 that Salinas Valley, known as the Salad Bowl of the World, had been implicated in an *E. coli* scare. In the 2006 outbreak, the exact source of the bacteria was never identified by the FDA, so the organic Earthbound Farm was eventually cleared of any connection. Possible sources of contamination were the presence of wild pigs around the field or runoff from nearby cattle farms. Natural Selection Foods has since established a stringent safety testing process at all of its greens-processing facilities. However, it has made environmentally aware consumers think twice about eating greens grown by growers in a distant state and transported long distances in refrigerated trucks.

Food miles is another new term that has emerged out of concern for understanding the environmental impact of the distance food travels from farm to plate. It is a natural extension of figuring a "carbon footprint," which is a calculation of a household's annual energy use converted into tons of CO_2. Considering that fresh produce in the United States travels an average of 1,500 miles, eating locally is one way to leave a smaller carbon footprint.

Earthbound Farm has also come under scrutiny by supporters of local food for its impact on climate change. While based in the Salinas Valley, the farm's crops are grown on 40,000 acres in five Western states. Earthbound Farm greens are ubiquitous; as of 2006 they were distributed to three out of four supermarkets nationally, according to Samuel Fromartz, author of *Organic, Inc.* Processed in a 205,000-square-foot plant, lettuce is "chilled to thirty-six degrees, beginning a cold chain that will continue as the salad is washed and bagged and sent to supermarkets around the country, where it is sold within seventeen days." The environmental impact of picking up a bag of these pristine prewashed spring greens at the supermarket is high: not only is the lettuce trucked across the country, but it is chilled all the way. That's a lot of fossil fuel burned to deliver a bag of organic lettuce.

On their website, the owners of Earthbound Farm rightfully counter that they are keeping over 457,000 pounds of pesticides out of the environment annually, as well as conserving over two million gallons of petroleum by avoiding use of petroleum-based fertilizers and pesticides. They take environmental stewardship seriously, employing a variety of conservation measures, using renewable energy and buying greenhouse gas offsets. Still, the greens have taken a long and complicated journey from field to grocery store.

Ben March has picked up on the trend toward local foods. In the farm store, a large sign by the produce reads: "The average meal in the United States is transported 1,200 miles. Buying locally helps protect the environment and supports local economy." He tells me that the tender late-winter lettuce grown in a March Farm greenhouse is fertilized with a mix of nitrogen, potassium, and Epsom salts and, once picked and bagged, is transported within hours across the street to the farm store cooler. There has never been a food safety problem with this lettuce, but if there were, it would be easy to isolate and identify.

For many consumers, local foods, perhaps not strictly organic but grown by an area farmer, have become preferable to an organic product shipped across the country. The organic food movement was first called the natural foods movement when it emerged in the marketplace in the sixties. The basis of the movement, which goes back several decades before the sixties, is that food raised without chemicals was better for humans and better for the earth. Hands-in-the-soil small farmers and kitchen artisans were the early producers of this food.

As consumer demand for natural foods grew over the ensuing decades, organic products gained brand-name recognition: Cascadian Farm, The Hain Celestial Group, Odwalla, and Seeds of Change. Since 1997 large food conglomerates have acquired many of these smaller companies, companies originally founded in thoughtful environmental and health creeds. The acquisitions happened quietly, and the parent companies are often not listed on the brand-name food packaging.

General Mills owns Cascadian Farm and Muir Glen. Seeds of Change is owned by Mars, purveyors of chocolate. The Hail Celestial Group owns over thirty organic brands, including Hain, Imagine, Arrowhead Mills, and Celestial Seasonings. Odwalla, producer of healthy juices and nutrition bars, is parented by Coca-Cola.

To some consumers of organic food, a corporate umbrella can be a turnoff. Part of the reason people are turning to local foods is because they feel more connected to the maker of the food, something that was inherent in the early natural foods movement. However, buying local does not mean a food is necessarily organic. Perhaps the food doesn't carry a certified organic label, but it is still grown without the use of chemicals; or perhaps a judicious amount of chemicals are used only as necessary, as in the case of eco-apples grown in New England; or perhaps the food is grown conventionally, but it means a farm family can continue to own and operate their farm. Whatever the reasons, the interest in local foods is increasing annually.

Although Connecticut is a small state, and the loss of farmland has been happening at an alarming rate, consumers who have a little extra time and money have many options for finding locally grown food, either directly from a farm or farmer or indirectly, from retailers of farm food. Direct purchasing includes buying from a farm store, picking your own fruit at a PYO farm, buying from a farmers market, or participating in CSA. Buying indirectly means purchasing produce from enlightened grocery stores that carry local produce or patronizing health food stores carrying primarily organic produce, with an emphasis on local sources. As of the summer of 2010 Connecticut offered direct sales through more than 100 farmers markets, 56 CSA farms, and 74 farms featuring PYO fruit.

Connecticut even has its own quarterly magazine, *Edible Nutmeg*, which chronicles and celebrates local foods of the Nutmeg State. Started in the fall of 2006, editors Robert Lockhart and Mary Adams turn out beautiful magazines chock-full of articles and photographs on local farmers and foods, as well as directories of Farmers markets, CSA farms, PYO opportunities, and recipes for seasonal foods. The magazine is one of sixty-five across the country that carry the "Edible" name, all linked to Edible Communities, a publishing and information service that is the seed for the start-up of the regional and local publications.

Organic Food Owners

More information on ownership of organic food companies can be found at: www.msu.edu/~howardp/

The author of the website, Philip Howard, PhD, is an associate professor in the Department of Community, Agriculture, Recreation and Resource Studies at Michigan State University.

The CT DOA sponsors the Connecticut Grown program, which produces publications promoting local food products, as well as provide farmers with marketing support, such as the brown-and-white road signs directing customers to local producers and signs for produce bins. Another source of information is the buyCTgrown program, sponsored by CitySeed in New Haven and an advisory committee, which provides technical and marketing assistance to producers of local food. Website visitors can sign up for email updates on events and seasonal availability of Connecticut crops.

March Farm produce can be bought through a variety of outlets. Like most farm stores, the Marches' market is open seven days a week during the growing season. Five PYO crops are grown: blueberries, strawberries, cherries, apples, and peaches. They participate in Saturday morning farmers markets, as well as selling corn to area farm stands for indirect sales. They also sell to grocery stores, which identify their produce as grown at a four-generation family farm.

In 2007, Ben March changed the Connecticut Grown signs displayed at Bishops Orchards Farm Market in Guilford to a CT Farm Fresh sign identifying the produce as March Farm's. Tomato sales increased an average of 30 percent weekly, with some weeks as high as 50 percent. Other grocery stores, like Palmer's in southern Fairfield County, highlight their local farm produce in commercials. As Fromartz, author of *Organic, Inc.*, writes, "Where food comes from, who grows and processes it, and what happened to people and the environment along the way can bestow attributes that make it extra appealing."

There is something intangible yet real that is added to my eating experience when I know the producer of my food. The Eat Local Challenge website lists ten reasons to purchase locally (see next page). Among them are supporting the local economy, preserving open space, fresher and safer food, and tuning into seasonal eating. Most important, buying local means local farms stay in business. Without our small farms, all the other reasons would be irrelevant.

10 Reasons to Eat Local Food

▶ **Eating local means more for the local economy.** According to a study by the New Economics Foundation in London, a dollar spent locally generates twice as much income for the local economy. When businesses are not owned locally, money leaves the community at every transaction.

▶ **Locally grown produce is fresher.** While produce that is purchased in the supermarket or a big-box store has been in transit or cold-stored for days or weeks, produce that you purchase at your local farmers market has often been picked within 24 hours of your purchase. This freshness not only affects the taste of your food, but the nutritional value, which declines with time.

▶ **Local food just plain tastes better.** Ever tried a tomato that was picked within 24 hours? 'Nuff said.

▶ **Locally grown fruits and vegetables have longer to ripen.** Because the produce will be handled less, locally grown fruit does not have to be "rugged" or to stand up to the rigors of shipping. This means that you are going to be getting peaches so ripe that they fall apart as you eat them, figs that would have been smashed to bits if they were sold using traditional methods, and melons that were allowed to ripen until the last possible minute on the vine.

▶ **Eating local is better for air quality and pollution than eating organic.** In a March 2005 study by the journal Food Policy, it was found that the miles that organic food often travels to our plate creates environmental damage that outweighs the benefit of buying organic.

▶ **Buying local food keeps us in touch with the seasons.** By eating with the seasons, we are eating foods when they are at their peak taste, are the most abundant, and the least expensive.

▶ **Buying locally grown food is fodder for a wonderful story.** Whether it's the farmer who brings apples to market or the baker who makes bread, knowing part of the story about your food is such a powerful part of enjoying a meal.

▶ **Eating local protects us from bioterrorism.** Food with less distance to travel from farm to plate has less susceptibility to harmful contamination.

▶ **Local food translates to more variety.** When a farmer is producing food that will not travel a long distance, will have a shorter shelf life, and does not have a high-yield demand, the farmer is free to try small crops of various fruits and vegetables that would probably never make it to a large supermarket. Supermarkets are interested in selling "name brand" fruit: Romaine Lettuce, Red Delicious Apples, Russet Potatoes. Local producers often play with their crops from year to year, trying out Little Gem Lettuce, Senshu Apples, and Chieftain Potatoes.

▶ **Supporting local providers supports responsible land development.** When you buy local, you give those with local open space — farms and pastures — an economic reason to stay open and undeveloped.

Used with permission of the author, Jennifer Maiser.

Visit the Eat Local Challenge website to learn more:
www.eatlocalchallenge.com

CHAPTER TWO

Winter: Preparation

History of March Farm

The orchards fill the eye with beauty every season. In winter the orderly rows of trees, like pen and ink scratches on the snow, progress neatly across the hills. In spring these skeletons of harvest's promise leaf out into a watercolor wash of pale green; then blossoms on the apple and peach trees appear, creating an impressionistic haze. After the fruit has set, orbs of color dot the trees like a pointillist painting. Then the brilliant paint box colors of autumn — deep gold, orange, and ruby — seem to bleed away, as if draining into the soil.

The earth rolls in folds toward the distant misty blue hills, the farm buildings nestling into the shoulders of the orchards. Sitting atop an undulating topography from geological forces aeons ago, Bethlehem benefited from the later Ice Age's deposits of glacial till (ground-up rocks), which would become prime farmland soil.

Connecticut is an intriguing state to geologists. Its 5,544 square miles make it the 48th smallest state, ahead of only Delaware and Rhode Island, yet its borders contain four distinct geological areas: Eastern Uplands, Central Valley, Coastal Slope, and Western Uplands. The Western Uplands are divided into two regions, the Northwest Highlands and the Southwest Hills, where Bethlehem is located. In spite of the rocky soil that perennially spews stones to be removed by farmers to the side of the fields, resulting in the area's characteristic stone walls, the Western Upland soils are rich in nutrients.

A map of Bethlehem reveals an abundance of prime farmland soil, a term used by the U.S. Department of Agriculture to describe land that has the "soil quality, growing season, and moisture supply needed to economically produce sustained high yields of crops when treated and managed according to modern farming methods." As early as 1756, settlers took advantage of this soil to plant orchards, grow grains, and graze livestock. Land was the basis of an agricultural economy, and land ownership provided social status as well as the potential for economic advancement.

Bethlehem was originally part of Woodbury, its southern neighbor, which was chartered in 1659. Bethlem, as it was then called, means "house of bread." After receiving winter privileges in 1738, the people of Bethlem petitioned to receive society privileges, making a separate parish. The town was officially incorporated in 1787. Early

Above: The original farmhouse pictured in the 1940s. This house was demolished and replaced with the current house in 1974. Opposite: The farm in winter.

industry included grist, cloth, and paper mills; tanneries; blacksmithies and ice ponds, which became important with the increase of dairy farms and were in use until 1938.

Thomas Marchukaitis's landscape was drastically different from that of his grandson, Tom March. Michael Bell in *The Face of Connecticut* writes, "The Connecticut we see today is two-thirds forest land, one-sixth built-up land, and just one-sixth farmland. But in the mid-1800's, farmland covered three-fourths of the state. In place of the vast blanket of mature forest that native Americans knew and the extensive young forests we see today, Connecticut's landscape was a patchwork quilt of field after field, only occasionally interrupted by woods."

By 1930 the town's population was 544. The 1935 state census reported 35,000 farms, with twenty-nine percent of the population living rurally. Today 3,500 people reside in Bethlehem, and Connecticut's farms total a little over 4,000.

Thomas Marchukaitis arrived from Lithuania in July 1912 with his two eldest daughters. Their boat trip probably cost between eleven and fifteen dollars apiece. He left work as a farmer and forest ranger. A few months later, his wife, Rose Mazaika, and their six other children joined them. That same year, Matt, his ninth child, was born.

According to family lore, Thomas left Lithuania to avoid being drafted into the czar's army. Between 1868 and 1914, nearly 20 percent of the population of 635,000 fled Lithuania in order to escape political oppression. Thomas initially worked in Waterbury for Scovill Manufacturing, one of the dominant manufacturers in the Brass City. By the 1920s, more than a third of the brass manufactured in the United States, including war munitions, was made in the area.

Thomas didn't like city life. His wife, Rose, had relatives who had already settled in Bethlehem, a bit farther down Munger Lane. In 1915 Thomas and Rose bought the original 114 acres of the farm for $2,500 from

Thomas Marchukaitis, right, with his youngest son, Matt.

Name	Birth Date	Place of Birth
Thomas Marchukaitis	1860	Lithuania
Matt (Matthew) Marchukaitis (later changed to March)	1912	Unknown
Tom (Thomas) March	1944	Bethlehem, Connecticut
Tom (Thomas) March, Jr.	1976	Bethlehem, Connecticut

Left: Matt March clearing roads of snow, 1930s. Right: Matt in 1960s collecting eggs. The chicken house is now the apple house, where apples are cleaned and sorted and cider is made.

Nathan Hurd Bloss, whose family is attributed with establishing the first commercial orchards in the area, although Native Americans were the first known to grow apples. Rose served the community as a midwife.

The original Marchukaitis farm was primarily apple orchards, which every homestead had. The farm also produced vegetables, eggs, and milk and was run on manual labor and horsepower until 1935, when electricity was delivered to the poles at the top of the hill.

Nine children, including Matt, Tom's father, were raised on the farm. In 1937, Matt married another Bethlehem resident, Anastasia Skeltis. The union of two Lithuanian families was celebrated with the wedding of a decade. According to Bethlehem resident Dave Kacerguis, author of an informal history of the Lithuanians in Bethlehem, "The party began immediately in the morning after the ceremony and lasted all afternoon. You would think that it was now over and everyone would go home. Well, that was partly right. They did all go home but only to milk the cows because the cows don't care about weddings. When the chores on the farms were complete, the party began again and lasted into the early morning hours."

In 1928, at age 16, Matt took over the farm. He constructed all the farm buildings himself, including, in the late forties, the large cow barn behind the main farmhouse. Although no longer inhabited by dairy cows, the

Tom March at age 14, entering his cow Mooch in the Bethlehem Agricultural Fair. Tom and Mooch took first prize.

barn's ground floor is still used for storage. The renovated hayloft is now the home of Tom March, Jr. Below the farmhouse, two chicken houses with a garage were built; they are now the apple house and cider house. Thirty-six acres across the street were bought in two stages, fourteen in 1947 and twenty-two in 1958, bringing the farm to its current holding of 150 acres. The store across the street was built in 1956, originally intended for apple storage. Greenhouses were added starting in 1988.

Matt and Anastasia had five children, three boys and two girls. Matt passed on his farmer's blood to his middle child, Tom. Born in 1944, he worked on the farm from a young age alongside his father. The Marchukaitis name was legally changed to March in 1954 to keep billing and legal documents simpler.

Matt died in 1992, at the age of eighty, after a yearlong battle with cancer. By the time he died, Tom and Sue had been running the farm for sixteen years. Anastasia lived five more years in the house above the blueberry orchard, where Heather Hurley, Tom and Sue's oldest daughter and her husband, Bill, and their family now reside.

From the early twentieth century Bethlehem was gradually populated by Lithuanian immigrants with distinctive names like Welicaitis, Kacerguis, Butkus, Majauskas, and Assard, names you still hear in town today. By the 1950s there were at least 25 farms in town owned by Lithuanian farmers.

Thursday was the day farmers took their goods to the market in the closest large town. For Bethlehem growers, this meant an eleven-mile trip to nearby Waterbury. Tom remembers arriving at Green Street by six a.m., a couple of hours before school. The Marches always had something to sell, including the carp they'd catch in the stream that runs from Long Meadow Pond. "We'd ice it," Tom said, "and bring it to Green Street, along with the vegetables and fruit and eggs." He often helped out with the weekly egg route of 110 stops.

But Tom was never allowed to stay out of school on these days. He laughs, remembering, "Oh, no, you don't know how Lithuanians feel about education." When the early Lithuanians arrived here, they came from a country that had no public schools. They eagerly took advantage of the American public education. In 1917, Tom's great-uncle, Joseph Marchukaitis, was the first Lithuanian to graduate from Bethlehem's public school system, even though he had arrived in the United States just five years earlier. In 1944 Tom's father, Matt, was the first Lithuanian elected to the local Board of Education.

When Tom was growing up, Sunday dinners were crowded with relatives, who looked forward to a hearty meal. Ten to fifteen people would be around the table each week. Tom attended the University of Connecticut campus in Storrs, where he studied for two years in the College of Agriculture with a focus in dairy.

Tom and Sue Collett met through Sue's roommate. On their first date Sue told him, "I want to come up and milk a cow." Growing up in Meriden, Sue only saw cows over the fence at a nearby farm. Her parents teased her about the farm poster that hung on her childhood bedroom wall. The young couple married in 1973 and three years later took over the farm from Matt and Anastasia.

Sue's younger brother, Billy Collett, began working at the farm during the summers when he was in high school and liked the work. In 1979 he began working full-time. He left a few years after that to work at other jobs, but returned because he preferred farmwork. He keeps all the machines and tractors running and pitches in whenever any other work is needed. According to Tom, "He is indispensable. A real fix-it man. There is nothing he cannot do." Billy attended technical high school and has taught himself how to keep everything mechanical running on the farm. Now he and his wife live next door to Tom and Sue.

Their son, Shane, began working at the farm after graduating from Nonnewaug High School, which has a strong agricultural program. Shane has a passion for antique tractors. His favorite part of farming is being on a tractor, out by himself, seeing the harrowed field when it's done. For many years, Tom, Billy, and Shane kept the farm running, and Sue managed the finances, as she continues to do.

Aerial photo of the farm taken in 1999.

Until the late eighties, March Farm operated primarily as a dairy farm. At its peak, livestock numbered 100 dairy cattle, including 40 to 50 milkers, as well as 600 chickens. Orchards yielded apples, peaches, and cherries, as well as plums and pears for home use. In 1972 blueberries were introduced. In 1988, the dairy portion was shut down.

"Billy, Sue, and I made the decision, although my father had reservations," Tom says. "The fact is, we weren't making any money, and nobody liked the responsibility of milking cows twice a day. Milk prices were low, and everything involved in taking care of the cows — feed and fertilizer — was costly. You kept going backwards instead of forwards. The worry was constant, a knot in my stomach. And I never got enough sleep. It was such a relief to get out."

Tom smiles. "A lot of people in the county thought we wouldn't make it without the dairy, and there were a couple of lean years. But we're still here."

The chickens were kept until 1992. After that, "there was a lot of chicken soup and fricassee."

March Farm started growing greenhouse tomatoes in 1988. The farm also raises corn, apples, peaches, blueberries, strawberries, and cherries, and the family continues to expand their crops. Diversification is the key to making the farm a successful operation. If one crop doesn't do well in any given year, the others carry them through.

All four March children left home to attend college. Ben, their third child, worked in the corporate world for a few years before deciding to return to the farm in 2004. "A happy day," Tom says. The other three children,

On an October day in 2008, the March family gathers on the hill above the apple orchards for a portrait. Tom helps Sue onto the seat of the tractor. "C'mon, Susan, it's our time to shine," he says. He's not just talking about the two of them. The extended March family, from left: Ben March; Billy and Shane Collett; Heather and Bill Hurley; Emily, Treyden, and Tom Medonis; Sue and Tom March; and Tom March, Jr.

Heather, Tom Jr., and Emily, eventually returned to the farm to live. Ben works full-time on the farm, while the others hold outside jobs. Tom tells me that before Ben returned to the farm, he was "outnumbered three to one by Sue's family, the Colletts." More important, the odds are now more in favor of the farm continuing.

Growing Up on the Farm

HEATHER MARCH HURLEY

The oldest of the March children, Heather is a slim, dark-haired version of her mother, Sue, with the same lovely voice and the same thoughtful way of answering questions. Her memories of growing up are very happy.

The farm was a magnet for friends. "We had quads (all-terrain vehicles), we had snowmobiles in the winter. In the summer my best friend and I would use the lawn mower and cut paths through the grass and then we'd ride our bikes on those miles of paths. The tag games at night were enormous. We'd skate in the winter on the pond and have pond parties with a ton of people."

But there were things that were not so pleasant. "I can remember watching them slaughter the animals, the milking cows, beef, pigs, right in front of the big barn. It was the only time my mom would allow me to stay over my friend's house on a weeknight. I'd stay away for two days, until the smell was gone. Everyone would say, 'What are you talking about? There's no smell of blood.' But yes, there is."

The chicken coop was another place fraught with possible danger. "When we went to collect eggs, I was afraid the chickens would peck at me, even though I knew if you just kicked your leg out, they would move. Sometimes we would take an egg and throw it, and they would all swarm around, pecking at the yolk. That would clear the way." Her grandfather, Matt March, heard about this trick. "He was so mad. He said, 'You don't waste an egg; you don't waste anything.' I was trying not to cry. 'You're lucky you're not my kid. If I found your father doing this, I would have brought him behind the woodshed.' I don't know if I ever threw an egg after that."

Horses were a big part of her young life. In the hot months, friends brought their horses over to ride and then cool them off with a swim in the farm pond. "Taking care of the horses took a good part of my life until middle school. Then I went through that horrible phase where I didn't want to do anything, including taking care of horses, anymore." When her mother gave her the ultimatum, Heather said, "Fine, get rid of them." She adds, "To this day, I wish I hadn't had that attitude."

As the eldest child of a well-known farm in a small town, Heather was ever visible. "My whole life I've walked into the hardware store and charged everything because everyone knows who I am. I was afraid to sneak out or do anything wrong because my parents knew everyone and I would get found out." She smiles. "It was a real advantage to my parents because I was a really good kid."

"By high school, being on the farm wasn't fun anymore. I wanted to get away. I needed to leave."

Franklin Pierce College in New Hampshire, three hours away, provided enough distance for Heather to gain some breathing room. At first she came home infrequently and would return early to school over Christmas break to work. "It kept me away for a while and let me learn more about other places besides Bethlehem, Connecticut, and March Farm. My friends at school told me I was such a naive little farm girl when I first got there."

Then something shifted, and Heather's appreciation of the farm was reawakened. "Every summer all my college friends came here and camped out. They thought the farm was the greatest thing in the world. We went for

Silos next to the former dairy barn.

hayrides, and my dad brought us through the fields in the back of the truck. They nicknamed it Marchstock, like Woodstock."

A teacher now, Heather has come back to live on the farm with her husband, Bill, to raise their sons in the house above the upper blueberry orchards, where Matt and Anastasia spent the last decades of their lives. "I can remember the day my grandfather died. He was as alive as possible. He walked out to his garden like it was his last day. If you don't have faith in a higher being, it almost puts faith into you. Suddenly he knew he wasn't going to be here anymore, and he got up that day and went out to look at the farm. That night, he passed.

"After my grandmother died, a few years later, we rented the house out. As long as someone else was living in it, I got an eerie feeling whenever I was in it, but as soon as Bill and I bought it, I never got that feeling again. It was like my grandparents were now at peace that a March, their first granddaughter at that, was living in their house.

"With my brothers and sister back here, it's like our own commune. My kids will grow up here on the farm like I did. That's what matters."

Tom March, Jr.

Tom Jr., quietly handsome and soft-spoken, doesn't look like a daredevil. But when he talks about growing up on the farm, he says, "I'm surprised I'm still alive. One time, I was probably around ten, my mom pulled in the driveway and I was on top of the silo with my friend, walking around the rim. She just about had a heart attack. She waited until we got down before she started screaming."

Tom liked climbing so much, he brought it inside. On the ground floor of the former dairy barn, a rock climbing wall takes up one section of wall. Pegs for footholds and handholds are secured into the wall; piled-up gym mats cushion falls. Tom pursued both rock climbing and snowboarding on numerous trips out West.

"This place has given me the freedom of my imagination. If I could think of it, we could try to build it. That taught me a lot, through trial and error. My father never really tried to stop us if we had an idea, as long as it didn't screw up what he was doing. We've got welders and everything else in the shop. I built a houseboat once — two big inner tubes with a platform and a little house on it. I can remember not being able to get to sleep at night because I was so excited about getting back down to the pond. I hated going to school."

Like his Uncle Billy, Tom has a mechanic's gift. He designed and built a heat exchange system for the greenhouses that relies on an outdoor wood furnace instead of costly oil. When his brother, Ben, needed a short bridge over the spillway of the pond to complete a farm hike trail, Tom put together a sturdy, attractive wooden bridge in a few hours.

His creativity is also apparent in his own home, the second floor of the former dairy barn, where he lives with his fiancée, Lisa. In what used to be the hayloft is now the kitchen, draped in sheets of construction plastic. Tom has designed the renovation and done the work himself. The barn was used for storage starting in the late eighties, when the Marches decided to exit the dairy business. Tom looks around the kitchen and laughs. "Yeah, I guess I benefited from when the cows went out." He has added a gable that provides a view to the west from the master bedroom. "The sunsets are especially nice in the fall, with the colors reflected in the pond."

From this window he can see the house his grandparents lived in. "My grandfather dying was the biggest reality check for me. I was twelve or thirteen. He was the main man on the farm, and I was close to him. He never really scared me, like he did my siblings. Both my grandfather and father taught me about hard work. I remember seeing all the guys unload the hay and wanting to work with them when I was little. I used to climb up on the hay wagon and kick the bales down just so I could feel I was helping out."

Tom has worked for a local plumbing company for about ten years. He learned the trade on the job. "You get to know everything when you do plumbing, so I'm always doing something different." He pauses. "In late summer, I feel bad every morning when I leave for my job. I see them taking the corn off the wagon and counting it all up. It's the crappiest job, picking corn every morning. But I help out on Labor Day and when they get real busy. I do what I can whenever they ask me to."

Although he has done his share of traveling, Tom is here to stay now. "I've been out West and across the country eight or ten times, been all over, but not more than a couple months at a time. I get homesick and end up coming back. Growing up, we had so much freedom. Everything we needed was here. It's probably why I'm still here.

"I think I'd like to come back and work on the farm eventually. I'd like my uncle's job. He fixes everything. That's my plan."

BEN MARCH

Ben's wide shoulders and muscular physique hold up well to farmwork, allowing him to move through his day with a steady solidity, like a steam engine. The third of the four March children, he graduated from college in 2002 with a degree in business, then worked for two years in the corporate world. Time spent in a cubicle calculating Citibank employee pensions was enough to cement his decision to return to the farm permanently.

"I think of farming as being in my blood," he says, waving his hand at the orchards on the hillside. "Who wouldn't want to do this? I had thought about staying on the farm before going to school. Working in the corporate environment helped me clarify my values."

Ben learned early that if he worked, he could make money. Sue and Tom March did not require the children to perform a long list of chores. Instead they could earn money by collecting eggs, picking blueberries, mowing lawns, or picking up drop apples for cider. Ben's entrepreneurial spirit emerged at a young age. He would spend an entire day selling lemonade to pick-your-own customers.

Lessons in capitalism were just part of his education. "There was no person in particular who taught me the most. The people I work with now aren't what people could consider typical farmers. They're all creative thinkers, so you get different points of view."

In a time when family farms are challenged, Ben is bringing innovation to the farm. His focus has been two-pronged: expanding markets and creating agritourism activities. Market expansion has occurred through widening the farm's wholesale presence and participating in local farmers markets. New agritourism endeavors include a corn maze, a hay bale playscape, and a farm hiking trail, all free of charge, with the intent of enhancing the farm experience and encouraging customers to spend more time at the farm.

Ben's own experience of the family farm is filled with sweet memories, including activities tied to every season. "In the spring we'd walk the entire river from the pond up to Long Meadow; in summer we'd be in the pond. There were hay forts and leaf piles in the fall and winter sledding anywhere and everywhere on the farm, with tracks built up for jumps. My siblings and cousins were always around."

When asked about his favorite smell, his answer differed from that of other members of the family, who had named hay as a heavenly scent. "The smell of pine is my favorite. I smell it when I'm not here, and I have an instant memory of the farm. I remember playing with a Wiffle ball in the backyard and thinking that the pine trees were enormously high. And the sound the wind makes, a soft swishing sound, that's my favorite sound, too."

The Connecticut state motto is "Qui Transtulit Sustinet" ("He Who Transplanted Still Sustains"). Ben's return to the farm is the reason the farm will continue for another generation; his commitment to the farm is the sustaining factor. Yet he understands the value of the experience of those before him, recognizing that "the best people on the farm are the ones who've been here the longest."

Ben now lives with his wife, Larissa, and their daughter in the house behind his parents'. He has found his place in the world here. "I feel my life has a purpose. I'm continuing something that was started a long time ago. If I can't say anything else about my life, I did my best to keep a business going and tried to turn it into something people know about and hold in high regard. I take a lot of pride in the farm."

EMILY MARCH MEDONIS

Emily, the youngest of the four March children, is slim, blonde, and athletic. Deep dimples crease her cheeks when she smiles. She is a runner and dedicated mother of three children, two boys and a girl. Like her siblings, she remembers her farm upbringing as happy.

"I have thousands of memories of growing up here. I remember sitting in the grain bins while my dad would feed the cows, and we'd go back and forth with him. The smell of tomatoes ripening, the hay in the barn, the apples in the coolers, the sounds of the peepers at night, the tractors in the morning — I love it all."

Being fourth in line shaped her personality. "I became more responsible because I wanted to grow up fast, to catch up to my siblings. I was always told what to do, so it made me pretty laid-back. Any youngest child becomes adaptable because they're forced into being a tagalong." It also made her independent. "I still get frustrated because I want to figure out things for myself."

Lessons of farm life were absorbed easily. "We learned a lot from what we saw. We were never forced to participate or told we had to do certain things. I learned to treat people with the respect, just like I would want to be treated."

Now that she has a family of her own, interests like canning and preserving —what she calls "old-fashioned skills" — have become stronger. "Every time there's something rotting in the store, I'll plan my whole meal about it."

Emily waitresses at a local restaurant and watches two other children four days a week. She credits growing up on the farm for her work ethic. "Hard work pays off. There's things you need, and there's things you want."

I comment on her framed pen and ink drawing of a barn hanging in Tom and Sue's house. "That was in high school. I remember going up on the hill when I was eleven and trying to draw the farm. I liked art and was pretty good at it. In college I was going to be a phys ed teacher, but I changed my major to graphic design." Emily designed the website for the farm. "In the future I would love to be involved on the farm, especially in the store, the number one marketing tool. The customer's first impression is what's going to make them come back."

She supports the expansion of agritourism at the farm. "I love how everything is free. It makes people feel welcome. A lot of people come just to use the playscape. With the farm hike and mazes, what family wouldn't want to come out here? Everything is so expensive now. It can cost $100 to plan a family activity. Here you come and pick blueberries and take part in the other activities and spend maybe $30."

Emily and her husband, Tom, live down the road from the farm store, in a house they bought and remodeled. When I ask what brought her back to the farm, she says, "I wouldn't have it any other way. I always knew I would come back. Living away from here, I wanted to be home. This place is such a part of me.

"I would love my kids to have what I grew up with: the freedom to be outside, the pond to swim in, a place to explore, the forts in the woods. It gives you such an appreciation of nature. My children don't watch a lot of TV, maybe an hour and a half a day. Kids are supposed to be out playing or, even when they're inside, using their imagination.

"I count my blessings every day. Not many people have this opportunity. It's incredible that all my siblings are here and the kids have their cousins (her sister Heather's children) up on the hill. It's the whole package here. It's freedom. I want that for my children."

SHANE COLLETT

There's no one at March Farm more passionate about tractors than Shane. At his recent wedding reception, he and his bride, Chantal, rolled into the party and were presented as husband and wife atop a John Deere.

The son of Billy Collett, Sue March's brother, Shane is one of the five full-time farmers at March Farm. He grew up in the house next door to Tom and Sue, at the corner of Munger and Bellamy Lanes. His playmates were his cousins, the four March children.

From a young age, the farm machines fascinated him. His favorite place: on a tractor. His favorite season: spring, when the plowing starts. His favorite activity: baling hay, if he's not in a hurry. And like many of the March family, his favorite smell is fresh-cut hay.

Whenever I see him, he's wearing overalls, a cap and heavy lace-up work boots. Medium built with light brown hair, he looks sturdy and strong. When he finishes his workday, it's not unusual for him to head to a nearby farm and help another farmer finish milking his cows.

His father taught him the most about farm life. "He took very good care of me and gave me what I needed. For the things I wanted, like a snowmobile or an all-terrain vehicle or, later, a tractor, I had to pick blueberries and do other chores to earn money. I learned to work for what I wanted."

Shane remembers fearing Matt March, Tom's father. "When we were breaking eggs in the chicken coop, I was always afraid he'd catch us. But I remember him going down the road on an old John Deere. It's no longer here on the farm" he adds. He takes down one of the miniature John Deere tractors that fill a shelf running around the top of the living room. "Look at the precision. Every part works on them. They even have serial numbers." When he was twenty, Shane had the green leaping deer John Deere logo inked on his arm.

He attended Nonnewaug High School, which serves Bethlehem and Woodbury. The school is well-known for its agriculture education program. Shane began working on the farm full-time after graduating. It took him a while to win over his uncle, Tom March. "Now I'm the same way with the high school kids when they start working here for the summer. You're lucky if you get a whole season out of them before they quit. I think a lot of kids come to a farm and expect to be driving a tractor, but that's not how it works. I did a lot of crappy jobs before I got to do that."

Shane and his wife and their young daughter live in the house he grew up in. His father and stepmother have built a house on the lot behind him, with a view of the western orchards. Like his March cousins, Shane is here to stay. The next generation has put down roots on the farm.

Winter Pruning

On any winter day, as long as it's not raining, it's easy to find Tom March. Look for a tractor up in one of the orchards and he'll be close by, pruning fruit trees. Late in the morning on a March day, the air is still nippy, and a heavy gray cloud mass fills the sky. Patches of blue emerge intermittently, but not long enough to generate any lasting warmth. Tom is dressed in a chamois plaid shirt over several layers, heavy jeans, a knit hat, and gloves. He holds a four-foot pruning wand, its long hose attached to the tractor for hydraulic power. He wears no safety glasses.

"Be careful," he says as he positions the pruner flush to the trunk of an apple tree. "I've had them hit me in my mouth." *Pfft, pfft, pfft,* goes the pruner, and branches fall to the ground.

The trees are planted fifteen feet apart and are trimmed to a width of seven and a half feet on either side. The lowest branches are supported by pieces of old furniture, bedposts, or table legs, to protect the tree from wind damage. Many of these trees are forty to fifty years old, past their life expectancy of thirty years, but still producing. The branches of the peach trees are straighter, and the shiny silver gray bark is banded with dark striations, making them look like pewter birch trees. New growth tops the trees with a claret red crown.

The pruner is powered by the tractor, which runs constantly. The clipping end of the wand resembles a toucan beak. Tom uses a small Stihl chain saw for the larger branches, handling it with the ease of a deli slicer. It takes him about five to ten minutes to prune a dwarf apple tree, compared with the forty-five to sixty minutes needed when the pruning was done without hydraulics and the trees were full-size. Tall tapered ladders were used then to reach the upper branches. Now Tom can complete the annual trimming of 5,000 trees from the ground.

"Pruning gives me something to do in the winter when it's cold and gloomy," Tom says. He works with an instinct honed over fifty years of experience. His first pruning wand was a child-sized one placed in his eight-year-old hands by his father, Matt.

Top: Tom March pruning peach trees.
Above: Two of the five farm tractors.

Pruning helps trees develop the proper form, yield better-quality fruit, and live longer. Dead, diseased, and broken branches are removed first; then the tree is trimmed with a basic shape in mind. Peach trees are kept open in the middle with more of a wineglass shape. An apple tree is cut back to the rough shape of a Christmas tree, with a central leader branch maintained and top branches trimmed shorter so the sun will reach the lower branches. This also allows air to move through the

tree, promoting rapid drying (which minimizes disease infection) and permitting complete coverage when the spraying is done. The cuts on the tree will heal over naturally and produce a thick raised oval border where the branch was attached.

When Tom's done with the apple trees, they resemble a line of soldiers with amputated limbs. The smaller branches are left on the ground and will be mulched in early spring. Large branches are tossed in the bucket of the tractor to be used for firewood.

As soon as the weather warms up and the snow has melted, Billy Collett, Sue March's brother who works full-time on the farm, operates a flail mower through the rows of trees. A set of rubber blades attached to the front of the tractor sweeps up the branches left on the ground from the pruning. They are chopped into a fine mulch by the chopper on the back and left on the ground in the grass between the rows.

Billy stops the machine to get out and check a fuel leak, then takes out his cell phone and calls his son to pick him up because the leak can cause damage to the tractor. I ask Billy how many tractors the farm has: five. And how many are running at the moment? Two.

Tom tests the soil annually for corn. He squats down and digs up a clump of grass. "Look," he says, clenching a clod of soil in his palm, then opening it to reveal chocolate brown till that crumbles after squeezing. "The rotted apples, the mulch, and the grass all fertilize the soil."

The sun is high now, though generating no more warmth, and Tom is ready to break for lunch. "At least apple trees aren't like cows," he tells me. "The cows could never wait. Trees can wait a day or so." And wait they do, until Tom, in his winter vigil, prunes every last one of them, just as he will sort, wash, and polish every piece of fruit they produce, with his own two hands.

Top: Shane Collett making cider. Above: Billy Collett, Sue March's brother.

Homemade Applesauce

This is an easy and satisfying way to use apples that are overripe or blemished. Quantities can be adjusted to taste and the number of apples on hand.

Put ½ cup of water in a pan and heat over medium heat. Add:

• 6 apples, chopped (peeled only where necessary)

• A handful of dried fruit: raisins (I prefer golden, but either will do), dried cranberries, chopped dried cherries, or any combination

• A generous amount of cinnamon

• A pinch of cloves

• Juice of half a lemon

Simmer until the apples are the consistency you like. To hasten cooking, put a lid on the pot as you cook.

Makes three cups. Variation: Try half apples and half pears

Delicious warm or cold, as a side dish for breakfast (great on pancakes), lunch or supper, especially with anything containing cheese. Also nice warm with a dollop of plain yogurt or, for a richer taste, sour cream.

Apple orchard freshly pruned.

Blizzard of 1978, Bethlehem Style

By Tom March

March Farm in winter.

In February 1978, New England was walloped by one of the biggest snowstorms of the century. In Connecticut the storm lasted nineteen hours and dropped over two feet of snow. Massachusetts and Rhode Island were even harder hit. Connecticut Governor Ella Grasso, after trekking the last few blocks to the State House on foot, closed the roads across the state so the Department of Transportation could clear them. The storm's incredible strength came from hurricane-level winds of approximately 65 miles per hour. Here is Tom March's account of the storm.

It began snowing Monday morning around nine a.m. By noon the wind and snow were really coming down. By three p.m. the blizzard had hit. Twenty feet from the house it was like a curtain had closed. At times you couldn't even see the dairy barn from the house. The wind was gusting from thirty to sixty miles per hour.

Around three we got a call from the Carlsons up on the corner of Double Hill Road saying that Phil Hajjar was stuck. I got the snowmobile out and headed there. Our road, Munger Lane, was blocked except for a narrow path. I picked up Phil and started up the street to his house.

The wind was so strong and the snow so heavy, it was impossible to see. I was just guessing where the road was most of the time. Once we got by the woods behind my parents' house, the going was a bit easier because the trees blocked the wind a little. I got Phil home and came back through the orchards below my parents' house.

I was afraid to cut across the open fields because I thought I'd get lost in the middle, so I followed the hedge-row, which at times was completely blurred with snow and wind. By this time the wind had to be at its peak. I was lucky because the wind was at my back, and that made it a bit easier to see.

While I was on the hill, I decided to drive over to the Smiths', whose property bordered ours, and see if Mr. Smith could get out. His wife, Willie, was already at our house waiting for him. She couldn't get home due to the blocked road.

I found him trying to keep his driveway open. Traveling by car was impossible, so I headed home where there were now three people stranded. I told them I'd try to get them home after milking the cows. Meanwhile the storm was roaring away with no letup in sight.

After milking, I loaded Willie Smith onto the sled, and away we went. We traveled along the pond dam, and I was able to burst through a drift and into the field. Most of the pasture was windblown, but the walls were drifted with six to eight feet of snow. The wind was blowing so hard I couldn't see. I knew the drift was coming but saw it too late! The machine bogged down, and we were stuck. Poor Willie was scared to death, holding on to me for dear life. I managed to pull the machine out, and we went forward again.

The field was windblown, but you couldn't see anything. I managed to see the lights on my father's house and headed toward them. It was impossible to see anything else. Once we got to my folks' house, the rest of the trip was easy. I dropped Willie off at her house and headed home, again with the wind to my back. Even though the road was blocked, I thought I could get through on my snowmobile.

The next half hour made me realize how people get stranded in a blizzard. I was going along fine and only had about 300 feet to get to the corner of Double Hill when I hit a soft drift that started to bury me. I cut the machine to the right and climbed up on the hard snow from the last storm. Here I sat for a moment wondering how I was going to get through. The wind and snow seemed to be trying to level everything in their path.

Then something happened that I had been dreading. As I went through the big drift, it completely covered my machine and got the drive belt wet. When I tried to go ahead, the machine moved very slowly, with no power. I got hung up on some soft snow but was able to pull the machine out.

I started out once more, and the skis hit a branch, which stopped me. I became a little concerned when the machine just stopped and the light went out. I realized the carburetor was sucking snow. Here I was, sitting on the top of a drift, with the wind blowing forty to fifty miles per hour, wondering why the hell I didn't go back the way I had come up instead of thinking I could get through on the road.

After a couple of quick pulls on the starter yielded nothing, somehow I managed to hit the kill switch, which shut the machine off. I pushed the button to release the kill switch, and it started right away.

Now I had my light back on with only one more drift to go through. The drive belt was still wet, which gave me only about half power. I almost got to the top of the drift when the machine started to bury itself again. This time I wasn't able to pull it out. I tried pushing it from the back, but the skis just dug in more. Finally with a do-or-die attitude, I braced my feet and pulled with all my strength to move the machine. There was no way I was going to leave the snowmobile. I had just bought it. With my last ounce of strength, I yanked the 250-pound machine free by pulling it on top of myself.

The rest was easy. Before long I was free of the drifts and heading home. One down, two more people to get home. I wasn't so lucky with the second person, Mrs. Tolles, who was quite a sight in a snowmobile suit. We started out, but it was impossible. I knew a 60-year-old woman wouldn't last if we were stranded. We had two houseguests that night.

When we woke the next day, the farm was covered in white drifts. At the apple house, it reached the top of the garage doors. At the corner of Double Hill, where I had nearly buried the snowmobile, a whipping wind had built the snow up nearly to the telephone lines. I counted my lucky stars that I had made it home the night before.

This account is used with Tom's permission.

CHAPTER THREE

Spring: Activity

Tom's Tomato Town

It all begins with a seed. In this case, about 7,000 of them. Early in January tiny tomato seeds, smaller than a pinhead, are planted, 1,300 at a time, in flats in the basement of the main house at March Farm. They sit next to the furnace, warmed by a round-the-clock grow lights. In ten to twelve days, slim stems emerge, bearing two tender leaves. The flats are then moved to a temporary home in the smaller greenhouse called the grow house, where they'll be transplanted into four-inch pots.

A greenhouse is like a spa for the plants. The seedlings are heated, watered, and fertilized — pampered, basically — until they're big enough to be moved to their permanent home, a five-gallon bag of soil and placed in one of the twelve large greenhouses that each hold 600 plants. This process continues every two weeks. By May all the production greenhouses are filled with tomato plants in full bloom.

"You bring in plants, you bring in problems," Tom March tells me of the reason they start their tomatoes from seed. Tomato production began here in 1988, as March Farm was getting out of the dairy business. A nearby farmer, George McCleary of Tara Farm in Watertown, was raising greenhouse tomatoes successfully, and Tom decided to try it at his farm.

Above: Tomato seedlings all start in the farmhouse basement. Notice the words etched in dust on the furnace. Opposite: View from the lower western apple orchard. You can see the greenhouse roofs to the left.

The Marches have found it's a business worth expanding. Greenhouse crops have the distinct advantage of growing in a controlled environment that can be protected from bad weather. Today annual production is close to 7,000 plants.

"That's a lot of suckering," Sue says, referring to the process of pinching off excess growth from the plants in order to maximize production. Suckering is labor-intensive and must be done by hand.

By March the grow house is filled with color. Next to the table of the youngest tomato seedlings sit trays of flowers. Ben March, son of Tom and Sue, proudly shows me one-foot-high zinnias with healthy buds, nasturtiums with leaves the size of silver dollars, and marigolds with bright yellow flower heads. His mother has started these from seed and will eventually use them in hanging pots to brighten up the farm stand entrance. Pots of spinach, parsley, basil, and squash line the floor in white five-gallon bags, to be sold in the farm stand early in the

A greenhouse in January (top) and in March (above).

season. Heads of lettuce are ready: romaine, red leaf, green leaf, Boston Bibb. Ben cuts me a generous bagful to take home. Greens this fresh are tender and sweet enough to eat with the lightest of dressings.

On an early April day, the greenhouse plastic, snug and tight over its curved frame, glistens in the warm sun. Sounds from radios drift in from two different houses. If I follow the country music, I'll find Tom Sr. seated on a low scooter cart pulling suckers off tomato plants. The jazz music coming out of the house close to the pond tells me where his son-in-law, also named Tom, is preparing a greenhouse for plants.

A peek inside this house reveals what looks like a science lab: long rows of white five-gallon "pots" filled with soil line a white plastic floor. From each pot hangs a black feeder line that delivers water and fertilizer.

I approach another greenhouse, one in full production. Opening the door, I enter a sea of green. The sharp fresh scent of tomato plant leaves and the lulling warmth make me drowsy. Leaves tickle my arms with their light hairs as I brush past them. These are Match tomatoes, a determinate (bush) variety from Holland. Determinate tomato seeds are bred so that the fruit ripens more closely together in the season than garden-variety plants.

The plants top out at over five feet. Clusters of small fruit hang on every plant, along with plenty of yellow blossoms. A large fan at the end of the house moves heated air over the plants. I warm up quickly and step outside to cool off. I see why Tom and Ben work in T-shirts. By the end of May, tomatoes will start being shipped to wholesalers and local restaurants, which will note the farm origin on their menus. Plants will go to Stew Leonard's, a regional grocery chain, as well as to garden centers in the area. By the end of the year, March Farm will sell close to 88,000 pounds of tomatoes and 4,700 tomato plants, in addition to sales in its farm store.

In the farm store, I select a half dozen tomatoes and take them home. In my kitchen I pile them in the ceramic bowl on the counter and admire their deep, warm color, then hold one up to my nose and inhale the promise of summer. Outside my window the leaves on the trees are just beginning to bud. I'll have to wait until August for my own plants to bear fruit.

Buying fresh local produce is the closest I get to picking it out of my own garden. By doing so, I support a local economy. I might be keeping a farmer in business. The pleasure factor is important: the taste of these fresh greenhouse tomatoes is a vast improvement over the anemic gas-ripened tomatoes I could buy in the grocery store now. And with a recent E.coli scare, I feel safer buying these tomatoes. Additionally, my purchase saves the pollution that food transported thousands of miles creates. Climate change is on my mind, and every small action I can take, I do so willingly.

Plus, when I slice a creamy tomato open, watch the juice run out of it, and smell its acidic pleasure, I know that the hands that cared for it belong to a neighbor I trust: a local farmer.

Cuke and Tomato Salad

This is quick and simple and one of my favorites. It is a staple at meals once tomatoes are ripe.

• Chop 4 ripe tomatoes into small chunks.

• Chop 2 cucumbers (peeled, if desired, either completely or in stripes) into small chunks

• Toss together with chopped parsley and scissored basil.

• Dress with olive oil and vinegar (I love rice vinegar)

• Season with fresh ground pepper and salt.

Easy Lunch Salad

This salad makes a lovely presentation and is a satisfying and healthy lunch. Serve with whole-grain bread on the side.

• 1 small cucumber (or ½ large), peeled in stripes and sliced

• 1 medium tomato, sliced into medium thick slices

• 6-8 olives (Kalamata, Niçoise, and/or green)

• 6-8 basil leaves, washed and dried

• 2 oz. feta cheese, crumbled, or 2 oz. fresh mozzarella

• 2 tsp. olive oil

• 2 tsp. vinegar (rice or balsamic)

• 2 T pesto (optional)

Choose an attractive plate. Arrange cucumbers, tomatoes, olives, and basil in groups. Sprinkle cheese over top; dress with oil and vinegar. Drizzle pesto over everything, if desired. This salad can either be cut up and tossed at the table or left intact for "nibbling." Makes one serving.

Top left: Greenhouse frame ready for plastic covering. Top right: Tom lays the second layer of plastic on the frame. Out of view is Billy Collett, unrolling plastic on the opposite side of the frame and taking the same precarious walk as Tom. Above, right to left: Ben March, Shane Collett, and Tom March load a roll of plastic onto frame.

A New Greenhouse

The greenhouse frame awaits a calm, still morning for the plastic covering to be placed on the skeleton. A quiet, windless morning is required to handle the long rolls of plastic.

At 6:30 on a chilly April morning, the men begin by loading the roll of plastic onto the temporary scaffolding by the greenhouse peak. Billy keeps edging the machine close to the greenhouse until the men can grab the roll. Note no one is wearing gloves.

"Longest 280 feet I'll ever walk," Tom says. Tom and Billy walk forward on the frame, unrolling the plastic. Billy says, "Why don't we just run the rest of the way?" Tom replies, "Go ahead. I'll go with you."

Strips studded with nails are already prepared. Silhouetted against the silvery plastic, the four men set to work quickly, securing the plastic to the frame. Their hammers sound out a staccato rhythm.

Jamaican Farmhands

Everything about Rocky, whose real name is Gladstone Watkin, is broad, from his muscular shoulders to his smile. That wide smile exposes gaps between every tooth, which only makes him look friendlier. His skin is the color of bittersweet chocolate. From Westmoreland, Jamaica, he's been at March Farm since 2004. At home his own small three-acre farm, where coconuts, pineapples, yams, bananas, and citrus are grown, is tended by his wife and eight children for the six months he works in the States.

Seasonal farmworkers from Jamaica have been part of March Farm since the fifties. They arrive in April and stay until the first of November, leaving behind farms and families in a country the size of Connecticut.

They're hired because they're not afraid to work hard. Tom says of Rocky, "I can't keep up with him."

One day in the greenhouse, I see a new face, a thin, light-skinned, middle-aged man.

"Hallo, fine," he says as he walks by.
"Who's that?" I ask Rocky.
"That's Shaggy, mon. Averill." He elongates the last syllable.
"Where'd he get that nickname?"
"From school."
"Everyone has a nickname?"
"Yeah, mon. A nickname is good."
"Does Tom have a nickname?"
"Uncle Tom."
"Does he like that?"
"No."
"What about Bill?"
"Uncle Bill."
"And Sue?"
"Aunt Sue."

Jamaican farmworkers fertilize the blueberry orchards. Rocky is driving.

I laugh and so does he. Sue told me they also call her Mama, and justifiably so, because she sees to all the farmworkers' needs while they're here. They have become part of the March family. At a recent Memorial Day party, one worker carried around one of the March grandchildren in the crook of his arm.

The Jamaicans do whatever is needed on the farm. When they first arrive, in April they are in the tomato greenhouses, pulling suckers off the plants, then staking and tying them to support wires dangling from the ceiling. Most of the spring is spent tending the thousands of young tomato plants.

Rocky Watkin ties tomato plants in the greenhouse.

Later, there is sorting and packing tomatoes for the wholesale market. In early summer they gather blueberries for sale in the farm store. In late summer long days are spent harvesting and bagging corn. In the fall it's time to pick apples and peaches. They usually work together and always appear happy. No matter how hot it is, they are often in long sleeves and jeans. Rocky tells me that although they look hot, once sweat moistens the clothing, the dampness keep them cool.

"Have you ever been to Jamaica?" he asks me. When I say no, he responds in a lilting accent that I would "really enjoy the blue pretty water."

He's been coming to the States to work for twenty years, since he was 32, cutting sugarcane in Florida and working in Virginia, New York, Connecticut, and Maine. I ask him if it's hard to leave his family for half the year. He smiles and says, "They're used to it, mon."

March Farm has been hiring seasonal farmworkers since 1951. Matt March, Tom's father, initially used workers only for apple season, starting around mid-September. Tobacco pickers in central Connecticut, once finished with that crop, would then move on to apples. From here they might continue to Florida to cut sugarcane.

Before the workers arrive, Sue handles the applications that must be filed annually so March Farm can hire H-2A workers, the designation given to the visa required by the U.S. Department of Labor for temporary agricultural workers. For an annual fee of $400, the Marches purchase membership in the New England Apple Council (NEAC), located in Goffstown, New Hampshire. The nonprofit organization assists members in the process of applying for seasonal workers.

As of 2009, 196 growers belonged to the NEAC. The council secured 1,500 workers for 1,700 positions throughout the area. In Connecticut an average of 400 to 500 H-2A workers are employed yearly. The Jamaican farmworker population in New England is large enough to support a liaison office in Enfield, Massachusetts, in addition to the national office in Washington, D.C.

Once the workers are under contract, they must be paid for at least seventy-five percent of the contracted days at what is known as the Adverse Effect Wage Rate (AEWR). The workers are paid the higher of either the AEWR, the prevailing rate for a given crop, or the minimum wage. The Marches paid the AEWR, $9.16 an hour, in 2009, when the minimum wage was $7.25. The workers are compensated for the hours they put in, with no overtime, although they may work more than 40 hours a week. Out of their wages they pay a nominal fee ($20 per week in 2009) for excellent comprehensive health insurance through a voluntary employee beneficiary association.

Between the full-time employees and the seasonal workers, it's not surprising that the biggest outlay of dollars on the farm is for wages. In addition to the guaranteed pay, there are annual fees paid for each seasonal worker. In 2009, it was $1,250 a worker.

I ask Rocky what he brings when he arrives from Jamaica for a six-month stay. "Shoes and clothes — nothing else," he says. "I have to pay for extra luggage, and even when I just bring what I need, I still end up paying." For the same reason, he doesn't bring back anything extra for his family. The money he makes here helps pay for his children's schooling and associated costs. Of his eight children, five are currently in school. Three have graduated from the tuition high school all his children will eventually attend.

"Do you take a few days at the end of the season for a little vacation, maybe in New York City?"

A puzzled look crosses his face before he answers. "No. We go straight home, rest a little bit, then start working on our own farms."

When asked why there is such a need for H-2A workers, Joe Young, the NEAC's executive director, says, "U.S. workers don't apply, and they don't stay. The work is hard physical labor, and it's seasonal. I grew up on an apple farm. Being an apple picker was not my career goal." Workers from Jamaica are eager to come here because the unemployment rate is high in their country. Although they can make a living at home, they can make money quicker here. Farmers themselves, they have the skills and stamina to handle the job. Most New England farms use the same workers from year to year, relying on the workers' familiarity not only with farming in general but with their specific operation.

The Jamaican workers at March Farm live in separate quarters above the eastern orchard with one of the nicer views in town. From their back door they look out over the apple orchards stretched out below them and covering the hillside that climbs up to the next ridge.

For decades their residence was a small cabin that slept three. Tom remembers, when he was eight, watching the building, a former pig barn, moved up the street on 8x8 lengths of lumber. It was then remodeled into living quarters. By 2008, with the ongoing expansion and diversification of farm crops, there was a need for four seasonal workers, instead of two or three. The cabin was replaced by a modular home that accommodates up to six workers.

One early spring day, before its demolition, Sue March showed me the little cabin. Inside is a kitchen and a large room with three single beds, neatly made up. The deep yellow color of the walls make it homey, cozier than it looks from the outside. Once the new modular home is inspected by local building and state housing inspectors, this cabin will be torn down.

We take a tour of the new quarters, which now sits in front of the cabin. Sue points out the cement stoop donated by a neighbor whose wife rides her horse through the farm. The home has one large center room, containing the kitchen and living room. On each end are one bedroom and a bath. Each bedroom can sleep three. Everything is carpeted, and painted; curtains have been hung.

I stop by to visit Rocky at lunchtime in the new quarters. A large island, with stools perched around it, occupies the center of the kitchen. Aluminum foil is taped over the backsplash behind the stove and over the hood. "To keep them clean," Desmond, another worker, tells me, as he munches cold chicken and watches a pot of rice on the stove.

Most of the time the farmhands cook for themselves, often chicken and pork, prepared in a variety of ways: curried, jerk, fried, steamed, stewed, and in soup. Sometimes they take a meal at

Rocky scoops corn to be be bagged off the wagon.

a local restaurant. Unlike undocumented workers, they don't have to worry about being picked up by immigration authorities. Although theirs are the only dark-skinned faces in this small town of 3,500, they are comfortable mingling with white people. "People are nice here," Rocky tells me. "And where I live there are many different races, a real mix." They socialize with their Bethlehem neighbors and often use bikes to get around town.

Foreign-born seasonal farmworkers have been widely employed in the United States and in Connecticut since 1946. "Farmers have avoided the current H2A program because it is so sluggish. It currently supplies only about 75,000 foreign workers out of 1.2 million farm workers employed at peak harvest, or less than 2 percent," states a 2008 *New York Times* article. Since the total number of migrant and seasonal farmworkers in Connecticut ranges around 4,000, the number of H-2A workers — 400 to 500 — is higher than the national percentage of two percent. Hired farm workers make up a third of the total agricultural production in the United States, according to Department of Agriculture.

Tom Philpott, a North Carolina farmer who covers agricultural issues on www.grist.org, has written about farms in California moving their fields to Mexico to capitalize on the cheap labor. "U.S. farm employers can buy a whole day's worth of labor for a wage ($9.60) equal to an hour's worth of work at the going rate north of the border, while still doubling Mexico's minimum wage of $4.80 per day."

March Farm, far from the Mexican border, doesn't have that option and wouldn't take advantage of it if it did. The Marches have always gone through the appropriate channels, even though it means more paperwork and more money. It's not just the fact that if caught there could be trouble, including fines. It's the way farming is conducted here, following the rules, even if it isn't always easy.

On an autumn day, three Jamaicans are picking apples, working one tree. The spaces between the trees are filled with laden branches. Two workers stand on pruning ladders, which narrow like a woman's waist at the top. The other is on the ground picking apples at arm's reach. Around their necks they wear sturdy buckets with canvas bottoms that open up when full of apples. When the canvas is released and the fruit falls into the wooden bins, it looks like the buckets are delivering litters. Each full bucket weighs around 35 pounds.

It is quiet out here in the orchard, but when the workers are bagging corn behind the farm stand, reggae is blasting out of a boom box. It was Rocky who suggested moving the corn sorting out of the field and into the plastic-covered greenhouse hoop, where it's cooler and closer to the farm stand.

"In Jamaica my favorite music is reggae, but here in the States it's funk," Rocky says, swaying a bit. "I cannot really dance, but I still like to move to the music " Then he turns back to work. My favorite photo of Rocky is one taken at the cornfield. He is up on the corn wagon, in a bright orange sleeveless shirt and yellow rain pants, pulling in ears of corn, intent, his arm muscles massive. When I give him a copy of the photo, the other workers look over his shoulder.

"You look buff," they tease him. When I ask him how many extra copies he'd like, he hesitates. A fellow worker suggests four. I offer six. Rocky nods. "My kids would love to see these."

We Were Due

In May 2007, a three-minute hailstorm nearly obliterates the Marches' budding apple orchards. Tom told me a few weeks before, pride in his voice, "We've got a beautiful crop this year." When I see him after the storm, he says, "My apple crop is ruined." At my gasp, he adds, "I've got insurance. I'll be calling the cider guys soon. They might not take the apples, though. The little nicks on the fruit are already turning brown."

We are standing outside the apple house. The western orchards rise up in front of us across the farm pond, hills filled with 35 acres of fruit trees.

"When was the last time something like this happened?" I ask.

"Fifty-six." He shrugs and smiles. "We were due."

We were due. The words stay in my mind. Tom March takes the long view. He feels lucky it's taken five decades for this kind of disaster to happen again.

"I guess I'm an eternal optimist," he tells me.

I walk up into the orchards to look at the fruit. Deep divots disfigure the young apples, only two inches in diameter. The skin will either grow over the divot and leave a large brown blemish, or rot and ruin the entire apple. Baby peaches, like nicked fuzzy olives, show the same damage. You'd have to be an optimist to love this fruit.

Tom's three simple words — *we were due* — touch on the farmer's values, iconic in America, embedded in our collective psyche: persistence, hard work, self-sufficiency, independence, pragmatism, resilience. Farmers lead apparently simple lives that are culturally symbolic. These lives are containers for values passed down from generation to generation. As a culture, we look to the farmer to maintain these values for us.

As if they didn't have enough to do.

Tom March is a compact man in his sixties. His eyebrows are brown and wiry, his hair gray and thinning. He wears farmer clothes: heavy jeans and T-shirts, Carhartt jackets. He is usually in good spirits, ready to talk about the farm. He does have a temper, though.

One rainy November Sunday my husband and I stop by the farm store to pick up cider doughnuts. Tom stands outside, hands shoved in his jean pockets, watching rainwater run down the recently paved road straight into his blueberry orchard. Hay bales placed along the edge of the road abate the flow a bit, but the water, filled with the early winter mix of salt and sand applied the week before, at the first snowstorm, is quickly deepening ruts along the road.

"Why do they wait until November to pave the road when they had all summer? Now they won't be able to fix it until spring. All this salt and sand will keep running into my orchard." The ground is nearly frozen, so any drainage problem can't be dealt with until thaw. We stand inside the store, listen to him rant, then encourage him to call our First Selectman, which he reluctantly does.

"I'm not complaining," is the first thing he says on the phone. The head of the road crew arrives within a

Bee boxes are used in the orchards as well as in the greenhouses, to aid pollination.

half hour to assess the situation. Tom seems grateful for the response. My husband and I laugh later over his, "I'm not complaining." Exactly what he was doing — and rightfully so.

Farms are key to Bethlehem's charm. Although no one has ever defined the term *rural character* to unanimous accord, residents agree that this desirable trait is largely created through a landscape of rolling hills, fields, silos, and barns. The town covers over nineteen square miles in southern corner Litchfield County. Its population is a little over 3,500. About half of the residents are weekenders whose main homes are elsewhere, typically in New York City, about a two hours' drive south. Approximately 50 percent of Bethlehem is farmland or forest, with horse farms becoming more and more prevalent alongside the handful of remaining dairy farms.

Preservation of rural character was greatly enhanced by the purchase of the town's first piece of open space property, in 2003, the 123-acre Swendsen Farm on the east side of town. Sam Swendsen retired from the dairy business in the early nineties and began leasing the farm to Tom March to grow corn and hay. Tom was a key player in the conversations between the town and Sam that led to the purchase. As chair of the Bethlehem Conservation Commission, I was invited, along with others involved in the preservation effort, to tour the farm.

On a golden late afternoon in summer, we piled into a hay wagon pulled by a tractor, with Sam Swendsen at the wheel. We rode the entire perimeter of his farm, along the edges of cornfields, to the eastern side, where East Spring Brook runs at the bottom of a one-hundred-foot ravine. We ended up by the small pond where Sam watered his cows. We were all charmed by the idea of the town, with its rich agricultural history, purchasing a farm to preserve permanently.

Afterward I wrote a note to Sam, telling him I would to do anything I could to help save the farm. Half of the $1 million purchase price was covered by a state grant; the other half was to be bonded by the town, once it passed a referendum. Through a New Year's Eve dinner and auction, we were able to add $65,000 to the pot.

After the sale was final, the Conservation Commission assumed stewardship of the farm. Tom signed a new lease with the town to continue farming the land. This was the beginning of my ongoing interaction with Tom March.

One early winter day we drove over to Swendsen Farm in his pickup truck to look at the tree line. He wanted to trim it to eliminate shade over the corn crop. We walked the hedgerow, marking specific trees to save and those to cut down.

"I'll do whatever you people want me to," he said. "You people" referred to the environmentalists involved in land preservation. I could hear the small nugget of resignation in his voice.

Whenever he had a question or comment about Swendsen Farm, I'd hear the "you people" phrase. This continued for several years, until, one day, I noticed it had disappeared.

Somehow that felt like a compliment.

Every time I see Tom after the hailstorm, I ask him about the apple crop. He shrugs and says, "I'm waiting for what the insurance adjusters say. They come out in the fall."

Crop insurance covers disasters: drought, hail, frost, hurricanes, fire, excessive moisture, insects, plant disease, wildlife damage. Farmers are required to carry insurance by the federal government; if a disaster occurs anywhere in the country involving, for example, apples, then March Farm would be eligible for federal disaster relief for apples in the future, if needed. The Marches insure apples, corn, and peaches for weather damage.

I suppose "We were due" is the farmer's version of "Shit happens," but there is an essential difference. The latter implies a randomness to events. Tom's words speak to a sense of fairness in the inevitable downturns. Mother Nature's acts are not personal. She delivers them without favoritism.

Tom seems at peace with this. He has seen tough times and weathered them. Everyone has misfortune sooner or later in their lives. David Mas Masumoto, a third-generation California peach and grape farmer, writes in his book, *Epitaph for a Peach*, "A type of humility marks a real farmer. Those of us who battle nature all year must ultimately accept the hand we're dealt. We're cautious even at harvest, privately smiling when we discover that the cards we hold may be OK, inwardly grateful that there hasn't been a disaster. We hear of someone else with bad luck: a farm caught under a hailstorm, a plum orchard that bore no crop, a vineyard with a mildew outbreak. Success is relative. We pick our fruit and whisper to ourselves, 'It could have been worse.'"

As it turns out, Tom's optimism pays off. The insurance adjusters are generous with their appraisal. He is able to save more apples and peaches than anticipated. And the insurance company increases its estimate of what his orchard is worth in production value, which pleases him. Although it means a higher insurance premium, it also means he receives more insurance money when the next hailstorm hits in mid-June 2008.

So maybe, in the end, Tom gets what he's due, too.

Keeping the Farm Afloat

An interview with Sue March

Sue March is the bookkeeper for the farm as well as the chief baker. She serves the community as a volunteer on the Bethlehem Ambulance Association. When my husband had a kidney stone attack one afternoon, it was Sue who showed up, along with the rest of the team, decked out in her dark blue ambulance outfit. I could immediately see why she was on the crew. Her customary calm, easy manner was reassuring, just as her presence on the farm lends it a touch of grace. Tom says the farm has been able to prosper because of her.

Have you always kept the books, even when Matt and Anastasia were alive? We took over the farm in 1976, and that was the point where I also took over the books. Anastasia was doing it before. It was kind of learn as you go. I adopted what she did and made some shortcuts. They were alive quite a bit after that but had turned the farm as a business over to Tom at that point. We paid them a certain amount every month toward the purchase price, and all the bills and the books came here. I think it was a relief to her to have me take over, and it wasn't too long after that I started taking over their personal books, too. After Matt died, there was a lot of Medicare stuff, so I did that for her.

What do you like about being the bookkeeper? Dislike? I think it's important that somebody here has to know how much things are and where the money's going and how much money there is to spend. I'm the only one who really knows that, in essence. Everybody else here just likes to spend it. Tom is forever saying, "They said we don't have to worry about paying right away." It doesn't work that way. It just escalates. The good part is that somebody is keeping track of things. The bad thing is that I always have to be the bad cop and say, "Forget it — you've got to think of something else."

What has been the key to keeping your farm financially afloat? Keeping track of the bookwork and crop insurance. The federal government comes up with disaster programs at certain times. Without it, I don't know how any small farm could keep going. I've 'worked' here for a long time without paying myself when I was home for the kids. I didn't go back to teaching until after they were grown and gone. It was part-time work. Though I wasn't making that much, it helped a little. I had fifty percent benefits. Our health insurance is $937 a month now, and we pay it all ourselves.

Is the farm financially self-sufficient? Has it always been? Yes, but we have debt. It has increased recently with the additional Jamaicans and the activity across the street — a new well, expansion of the store. But I guess it's time. The biggest bulk of the money that goes out is for salaries. Ben (March), Shane (Collett), Bill (Collett), Tom (March), me — five full-time employees. Plus, the Jamaicans while they're here and other local people here in the summer. Then there's the teenagers in the store, plus Lynn (Horvath) and bakery help. Tom doesn't actually get a salary — he never has. Now he's collecting Social Security, but most of that goes back into the farm.

How are financial decisions made? We all get together and sit and talk about it. I'm usually the one putting the kibosh on things. We don't vote — we either have the money or we don't. Since Ben's been home, the expenses have gone up, but so has the income because we're offering more things. Hopefully we can keep the scales balanced. I think people enjoy coming out and being able to do something that's not breaking the bank. They can walk around and see the animals and do the playground and all that. And then they buy things, and that helps us.

Sue March in the commercial kitchen.

Has there been any point when you thought the farm wouldn't make it? No, not so far. Hopefully that continues. Once the well and the septic get done across the street, the next thing would be to do the store over. But I think that's a few years down the road. Ben's ultimate dream is to put the bakery over there, but I don't agree with that. I'm not particularly excited about having someone looking over my shoulder. When I'm here in the house, I can do other things. It was great when the kids were small because I could always tell by the amount of pounding on the ceiling, the foot traffic, what was going on. That worked out pretty well. Your ears get very sensitive.

What is the timing of planning? I think we should do more long-range planning. Ben and I talk about this, if we're both down in the greenhouse, for instance. Generally, because of the way the income flows, big line-item decisions are made in the summer or fall because that's when we have the money to pay for them. In the spring, we're just sort of scraping by.

What is your best hope for the farm? Your worst fear? My best hope is that future generations can continue with this and be able to have a viable lifestyle with it. My worst fear is that our debt load would get to a point where it would be impossible for people to continue and be able to pay it back. That's one of the things I always harp on to Tom. I see some of my children working here. Tommy has said for a number of years now, "When Uncle Billy retires, I want to take over for him." But the thing he has to understand is that he'd be taking a huge pay cut. I suppose there would be other benefits. He and Ben would work well together. They both have different skills, and I don't think they'd get in each other's way. Heather and Emmy both love living here. Heather's got her own career; her husband's got his own career. Emmy wants to do a day care thing. They probably wouldn't work on the farm at this point.

What is average farm income and expense? They end up being very close. Most everything that comes in goes right back out again. Enough said. It varies from year to year. As the income goes up, so do the expenses.

CHAPTER FOUR

Summer: Fullness

Blueberries

The berries are glazed with dew this morning. I come to pick early, when I can be alone in the quiet of the orchard, before the mid-July heat envelops the day. I carry a basket that can hold five pounds of berries. I wander the paths between the neatly mowed rows of bushes looking for berries ripened to a dusty midnight blue. It is a shade painters must strive for, the deepest indigo wrapped in a white veil. The color promises a taste explosion: sweet, juicy, and faintly tart.

Above: Blueberry blossoms before pollination. After pollination they will turn upward. Opposite: Cornfields are interspersed throughout the orchards.

I catch sight of a bush dipped in silver. The slant of the morning light strikes the berries as they stand on their stems, glistening like newly cast silver beads. For a moment, I have stepped into a painting created by a master's hand.

Yet as stunning as the hues are, berry color is not the truest indicator of ripeness. Instead it's determined by touch. Perfectly ripe berries do not need to be pulled off the stem. I roll the berries gently with my fingertips as I cup my hand. The berries that easily fall into my palm are ready; the others left to ripen. As I move from bush to bush, my fingertips searching the stems, I am content. Although I live only a half mile away, I feel completely removed from my daily world. I am in touch with something primal, the simple act of gathering food, and something modern, taking control of what I eat. It connects me to a natural rhythm of choosing nourishment directly from the source. The sun warms my head and arms, hinting at the heat to follow later in the day. A light breeze cools me briefly. I lift my head and take in the view across the apple orchard to the east and the smoky blue hills beyond. I smell freshly cut grass.

A familiar van pulls into the field. Tom March is driving. He pulls up next to me and leans out the window. "I was going to turn on the squawker, but you can cut it off if it bothers you." He's referring to a small box sitting in the middle of the orchard that emits sounds of a distressed bird to keep out berry-eating birds. Powered by a car battery, it uses a computer program for setting different birdcalls, which is necessary because in three or four days the birds figure it out. The sound is a short buzzy repetitive cry, but it's not intrusive.

Birds are the biggest threat to the blueberry crop. In the past Tom has used netting suspended on long poles to cover the lower, smaller orchard next to the farm stand. This orchard up on the hill is too large for that treatment. Several years ago he tried giant balloons painted with bulls-eyes suspended on poles. The

A former method of keeping birds out of the orchard: oversize "eyes" floating on long flexible poles. Today the farm uses a squawk box that imitates the call of a bird in distress.

huge eyes hanging above the orchard frightened the birds. The squawker has proved to be as effective as previous methods, although Tom is always on the lookout for a new way.

Blueberries are native to North America, and July is designated National Blueberry month. In New England, blueberry season runs about six weeks, from early July into August. Blueberries are cultivated in three categories: lowbush blueberries grow close to the ground; highbush blueberries can reach six to twelve feet; and rabbiteye blueberries, more suited for southern climes, are the tallest, at fifteen feet. Most northern orchard operations, including March Farm, use hardy high bush varieties. To prolong the picking season, March Farm grows four types, which are pruned to keep them at a good height for pick-your-own customers. True to their name, lighter-colored Earliblues are ready in the first weeks of July. Next the two mid-season berries, Bluecrop and Blueray ripen. The Coville, a darker, late season berry, is named for the government botanist whose experiments early in the twentieth century revealed the acidic soil conditions (pH 4 to 5) that blueberries require. Frederick Vernon Coville's work is the foundation of the modern strains of commercial berries grown today.

The berries emerge in mid-spring as clusters of slender acorn-shaped blossoms, either white, pale pink, red, or sometimes tinged pale green, which open to a bell shape. Once pollinated the blossoms point skyward, and a light green berry develops on the stem. As the berry grows and ripens to blue, the blossoms open and fall off. Although blueberries are self-pollinators, which means one type can be grown successfully by itself, they do best when more than one variety is grown close together.

Now my basket is full, but I linger. Time in the orchard is a quiet interval in the midst of a busy life. I absorb the peace and calm and store it away. Later today, or this week, when I remember the orchard, I will stop for a moment and breathe deeply.

My stomach growls. I think about pancakes studded with berries. I head to the farm stand and pay for my berries. On the way home I dip my hand repeatedly into the bundle of berries on the seat next to me and sample their luscious taste.

After a breakfast of blueberry pancakes, I prepare berries for the freezer. Blueberry pancakes, muffins, and cobblers are a welcome addition to winter menus. I put the berries in a big stainless steel colander and run cold water over them, while scooping from underneath with both hands, enjoying the pleasant feel of soft marbles. Once washed, I pick through the berries to remove any tiny stems or leaves. Then I spread them on paper towels, cover them with another layer of paper towels, and gently roll the fruit under the toweling to dry them off. I pull a piece of waxed paper long enough to line my baking sheet and lift the berries in handfuls, placing them onto the

Blueberry Cobbler

This is a good reason to freeze blueberries for winter use, or to run immediately to your kitchen after picking fresh ones.

Filling:
- 6 cups blueberries, fresh or frozen
- $^1/_3$ cup sugar
- 2 T cornstarch, flour, or arrowroot

Topping:
- 1-$^1/_2$ cups all-purpose flour
- 2 T sugar
- $^1/_2$ tsp. baking powder (I prefer aluminum free)
- 4 tsp. baking soda
- 6 T chilled butter, cut into small pieces
- $^2/_3$ cup plain yogurt

Preheat oven to 350. Combine filling ingredients in an 13 x 9 baking dish and stir until sugar and cornstarch are evenly distributed. Combine flour and next three topping ingredients in large bowl. Stir with whisk. Cut in butter with pastry blender or 2 knives until mixture is like coarse meal. Stir in yogurt to form a soft dough. Drop dough by spoonfuls onto filling. Bake for 40 to 50 minutes, or until blueberries are bubbling and top is lightly browned. Serve warm with vanilla ice cream or fresh whipped cream.

Ripe peaches in the early morning light.

waxed paper, periodically tapping the baking sheet of paper towels lightly to ensure the single layer that will facilitate even freezing. I pile more berries on, spread them out with my palms, then tap again. So satisfying, that tapping. When I finish, the sheet is filled with a single layer of blueberries, a mosaic of varying shades of blue nestled together in lustered color.

I stack the trays in the freezer. Later today, or tomorrow, I'll transfer the berries to freezer bags. This winter I will ration them out, holding off until after the deluge of holiday sweets to use them. The berries, chock full of antioxidants, will brighten our plates and our tongues in the muted months of winter, and remind me of more time in the orchard to come.

Orchard Fruits

If there was a contest for the most sensuous fruit on the farm, peaches would win first prize. In early spring young slender red branches, which turn pewter with age, are tipped with pussy willow-like buds. By May the small orchard, about 700 trees on five acres in the northern corner of the farm, appears swathed in pink tulle. Up close, a blossom's five pink petals open to reveal a cluster of green-tipped stamen. After petal fall in May, a small fuzzy olive-shaped fruit emerges, slightly yellow with a green tinge. As the fruit ripens, long elegant leaves drape and shadow the velvety buttons, which display a wash of sunset colors: yellow, orange, pink, gold, crimson. Close to ripeness, a silver gray branch yields clots of peaches, looking like a Joan Miró painting.

Eating a peach is a sense-filled act. First, sniff the peach for readiness. Once a sweet scent is evident, test the fruit by pressing gently with the thumb. A slight yielding should be apparent. Then the decision: dive right in, or delay and slice. The downside to diving right in is that sweet, tangy juice may drip down your chin onto your white shirt. But to experience the direct contact with the flesh of the fruit, the small tearing with your teeth, is to know the fruit in a biblical way. If you decide to slice, you might lose a bit of juice as it sheets the knife. Once sliced, lift directly into the bowl and cover with cream. This is real comfort food.

March Farm grows two kinds of peaches: yellow, which are tangy and sweet, and white, light and citrusy in taste. The skin of the yellow peaches is orangey yellow, while the skin of the white peaches is pale crimson and ivory. Even the names of the different varieties are easy on the tongue: White Lady, Cresthaven, Garnet Beauty, Canadian Harmony, Biscoe.

"People love the white peaches," Tom says. "Even the not-so-great ones, those hit by hail, are bought up anyway. I put them out and they're gone."

On a mid-July day, I head up to the peach orchard. The air is hot and dry, scented with the smell of grass cut a few days ago. The mowing goes on constantly in the summer, a soft background murmur.

After ten minutes of walking, sweat is trickling down my back. The distant hills to the west fade dark green to blue. A hawk screeches overhead as I approach. At the end of an apple row a stand of milkweed attracts monarch butterflies. My view takes in a cut hay field below the farm buildings nestled in their own little valley, up against the trees and the hills of the orchards. The hot air shimmers like desert air, making the farm look as if it's underwater.

I stop in front of a tree. The fruit is blushed to shades of the sun, the fuzz pleasing, the firmness just beginning to give, their scent slightly evident. I fill my basket and wander around the orchard a while longer.

Tom tells me that when picking peaches to be sold and transported, ripeness is indicated by color and the fruit needs to be hard or it will bruise when packed. Peaches used to be packed in tapered baskets that distributed the weight; now they are packed in layers in cushioned tomato boxes, protecting the delicate skin.

Second prize for sensory delight is strawberries, a fruit that carries its future on its skin. Tiny seeds texture the surface and feel like small bumps on your tongue. You have to work to pick the berries, bending down over the plants, pulling up the serrated leaves, which look as if someone took pinking shears to them. Underneath the leaves, the fruits hang like heavy heart-shaped lanterns. Unlike blueberries, picking strawberries is a two-handed operation. Hold the stem with one hand and tug slightly with the other to leave the rest of the plant and the less ripe berries intact. Three goats across the street in the petting zoo bray encouragement as I bend down. My knees crack, but the strain on the joints is worth it, for the fruit fills the basket quickly.

The sweet, clear flavor of strawberries is like a summer day on your tongue. Just the smell makes me salivate. As I reach under layers of leaves, an overly ripe one falls, forming a puddle of red mash on top of the wood chips. A yellow jacket finds it immediately.

Our first summer picnics include strawberry shortcake. I use the shortcake recipe from *Joy of Cooking*, consistently good. At our house we eat our shortcake rather decadently. Split a biscuit and put half in a good-sized bowl, the kind you might serve ice cream in when no one else is around. Throw on a dollop of homemade whipped cream, made from heavy cream and a bit of vanilla, sweetened with a long strand of maple syrup poured from the bottle as the mixer beats the cream. Taste the whipped cream only once for sweetness. It's pretty hard to get it wrong. Add sliced strawberries that have been sitting long enough in a little sugar to be juicy. A small bit of whipped cream before the top of the biscuit goes on, then more strawberries and one more spoon of whipped cream. Toss in a few

Storage Tips

- Don't wash strawberries until you're ready to use them. Moisture speeds up rotting.

- Strawberries are best at room temperature. Coldness dulls the flavor, so take them out of the fridge and allow them to come to room temperature before serving.

- Only cut up enough strawberries that you need at one time; they don't keep well once cut. If you have leftovers, freeze them for blending into smoothies. I keep a container in my freezer where I toss pieces of fruit unsuitable for eating but good enough to disappear into a smoothie.

- Peaches are ripe when their perfume is evident. Keep peaches at room temperature until fragrant, then store in the fridge.

- Cherries are also best at room temperature.

Young strawberries ready for picking.

blueberries if you want a patriotic look. Pick up a spoon and don't share with anyone, savoring the layering of sweet, juicy, creamy, salty, crusty textures and tastes. Enjoy every bite and hope there are more biscuits. If not, go straight for the strawberries and cream. This is not an indulgence to be shy about.

It's hard to imagine these luscious fruits in their embryonic state. Baby strawberry plants are just roots, cut short and rubber banded together in bunches of twenty-five stubby wet stems. Four plants are placed into a five-gallon white plastic pot. Forty rows of pots are lined up out on a black plastic ground cloth, which is then covered with wood chips. A thin feed line, looking like a plant IV, runs into each pot, delivering water and fertilizer to the plants. The source for this nourishment is a water tank and pump system that irrigates both the strawberry and blueberry patches. The water comes from the pond across the street.

The first round of berries should be ready in July, for these are ever-bearing (Everberry, Seascape, Evie2), with three ripening periods. In 2007, the Marches tried fifty plants; the following year it was increased to 5000. The current number is around 20,000. The tarnished plant bug is an insect that bites the strawberries after bloom and makes them "snarly." Elevating the plants off the ground helps control insect damage. The ground cloth helps keep the crop clean and eliminates nutrient-robbing weeds and grass from growing near them.

Although vulnerable to late frost, the berries overwinter well in the perforated bags, covered with sheets of white plastic. According to the suppliers, the plants last two to three years but Tom relies on the experience of his Jamaican workers, who tell him they've seen them last longer.

Cherries come in close behind strawberries in the contest of the senses. They are grown in a small stand in the upper western orchard, two long rows between the blueberry orchards and a field of corn. On a pick-your-own expedition I step under the tree into the branches' embrace. Looking up I see jewels: garnet, wine red, ruby globes hanging in clusters. A small tug releases the fruit, and I stretch to reach the higher branches, greedy for one more. The smooth slick skin in my mouth, rolled around in anticipation, then the break, the release of juice, deeply sweet, the soft meaty inside. The pit sucked, spit out. Then another.

Not long after this, I ate some cherries someone brought to a gathering. Thinking they were from a farm, they seemed extra delicious. When I found out that they were from a supermarket, they tasted the same, but the experience of eating changed. The grocery store cherries didn't have the same quality and care I imagined when I saw the entire scenario of the cherry acquisition — the buyer stopping at a farm stand, picking out the cherries with care, perhaps interacting with the farmer who had tended the orchard. This story made the burst of flavor juicier and sweeter. It reminds me of the premise from the *The Cluetrain Manifesto*, that today's consumer is looking for a connection that involves a relationship. How could that feeling about the cherries not affect how the cherries taste? And how could that not add to the quality of my life?

Farmers Market

The Litchfield Hills Farm-Fresh Market, as the colorful wooden sign at the entrance to Center School reads, is the "public face" of its umbrella organization, Litchfield Hills Food Systems, Inc. The idea for a market evolved out of the work of the Litchfield Economic Development Commission (EDC) during the ten-year update to the Plan of Conservation and Development in 2007. Subsequent grant funding support came from the Connecticut Department of Agriculture.

The weekly farmers market is held at Center Elementary School on Route 202 in Litchfield during the summer months.

Maintaining the rural character of Litchfield — a town of 8,500 that's a twenty minutes' drive north of Bethlehem — while encouraging economic development was the challenge put to the commission, and the idea of a farmers' market supporting sustainable and local agriculture was born. Once the mission of the EDC's challenge was expanded to include education, it was clear that a larger organization was needed. Litchfield Hills Food Systems, an educational services public charity, was created, with the mission of connecting sustainable agriculture, local foods, and active, healthy lifestyles.

"It's a three-legged stool," organizer and volunteer market master Kay Carroll explains. "Sustainable agriculture, healthy lifestyles, community. We are different from many markets because of our focus on education. Every week we have a chef demo. We have a Master Gardener. For kids we have a story hour and student art activity, centered on agriculture. In addition, we have guest musicians, guest artists, and space for local nonproducers." Managing all this means Kay puts in 50 volunteer hours a week.

On any Saturday morning between June and October, twenty local producers fill the parking lot of Center School, just south of the Litchfield Green. On a summer morning I head first to the nonprofit guest booth to purchase a basket I've seen other shoppers carrying. I choose a roomy, sturdy one with a deep turquoise and purple woven design and heavy leather handles. I browse the market. Each tent draws my eyes: bottles of golden olive oil in the sun invite the light; orchids hang in the air, their roots exposed; piquant homemade salsa and relishes are available for tasting; sample chunks of country French bread beckon from wooden bowls. My senses are in heaven. It is hard to linger at one booth without wanting to see, smell, and taste what's at the next.

Shoppers are dressed from fancy Sunday to Saturday casual. Some women wear heeled sandals and carry leather purses; others float by in long gauzy hippie dresses; shorts and T-shirts are the fallback fashion position. Guitar and violin music performed by Sweetheart Mountain add to the party atmosphere. In this relaxed setting, people are open and friendly, interacting easily with one another and the vendors.

At the March Farm tent, tomatoes are moving. A customer picks one up off the table.

"A real tomato," he says. "Not those red round things in the grocery store."

Also popular are chocolate chip cookies the size of a small dessert plate.

"Now, this is a cookie," one woman says, holding it up to show her friend. It's actually equivalent to three cookies. Crisp and buttery, with a judicious balance of cookie and chip, they remind me of the cookies that filled my Pennsylvania grandmother's green glass cookie jar. Although it's past prime strawberry season, March Farm is selling some from its thrice-bearing crop. I purchase one of the last cartons and begin planning dinner my favorite way, dessert first. Tonight it will be strawberry shortcake.

March Farm has been a vendor at this market since its inception in 2007. Although Ben was on the original steering committee and thought the market was a good idea for the area, he wasn't sure it was a good fit for March Farm, which has its own farm store. The Marches also had a previous, unsuccessful try at another farmers market. "We said, 'You really ought to try ours,' citing all the activities we had built in to be a destination," Kay Carroll says. "They did try it, and now they are thrilled. They can't bring enough products to sell."

Ben March arrives an hour before market opening to set up. With nonstop contained energy, he handles two shade tents single-handedly, lifts bags of corn and boxes of tomatoes, all the while answering my questions. I try to help but end up getting in the way. He wipes one of the white signs with Windex and is ready as the market opens at ten a.m.

"I haven't had good corn yet," a woman tells Ben as he bags a dozen ears for her.

"This will be best corn you've ever had," he assures her.

I count out a half dozen ears for grilled corn, a recipe my husband has recently perfected, turning out buttery tender corn with a hint of a smoky taste.

It seems every town I drive through these days has a sign for a farmers market. According to the CTDOA website, "Farmers' markets can be found seven days a week and just a short distance away from virtually any town." In Litchfield County alone there are nine markets; the 2010 season brought over 100 markets to the eight counties in Connecticut. This reflects the national trend: between 2000 and 2010 the number of operating farmers markets across the country more than doubled, from 2,863 to 6,132. The U.S. Department of Agriculture (USDA) website states, "Direct marketing of farm products through farmers markets continues to be an important sales outlet for agricultural producers nationwide."

Farmers markets are not just for the well-heeled. Most markets in Connecticut are affiliated with the Special Supplemental Nutrition Program for Women, Infants, and Children (WIC) and/or the Senior Farmers' Market Nutrition Program, which provide WIC clients and seniors access to fresh fruits and vegetables, benefiting not only those clients but also the farmers selling Connecticut-grown products.

The most recent USDA survey (2006) asked market managers to rank the top three reasons customers shopped at farmers markets. Freshness, taste, and access to local food were the leading responses, followed by support of local agriculture and variety. Nationally price ranked sixth, but in the Northeast it ranked lowest in seventh place. Nationally, freshness was rated as either important, very important, or extremely important by

98.3 percent of managers; in the Northeast, it was 99.4 percent.

I asked Kay Carroll whether these results jibed with her sense of the local market.

"As long as customers are not paying a premium, they'd rather have it fresh," she says. She studied the seven motivating factors in the survey. "Certainly all these apply, but they don't tell the whole story. This is the place to be on Saturday morning. It's the community gathering place."

The producers accepted for the Litchfield Hills Farm-Fresh Market have to meet certain criteria. Kay says, "All vendors have to be able to explain how whatever they're selling is made, and educate as well. We're picky about who we accept, and our vendors appreciate it."

The framework of the selection is written into the market's mission statement: "Our foundation stones are community-based programs focused on sustainable, local food systems that are respectful of the physical environment in which we exist. Our aim is to cultivate greater community by nurturing and supporting collaborative teaching and learning experiences, which connect and support sustainable agriculture, local food and active, healthy lifestyles."

Emily March Medonis takes care of an early customer on a chilly October morning.

Sustainable agriculture encompasses a wider range of agriculture than organic farming. According to Sustainable Table's comprehensive dictionary (www.sustainabletable.org), it is "a way of raising food that is healthy for consumers and animals, does not harm the environment, is humane for workers, respects animals, provides a fair wage to the farmer, and supports and enhances rural communities." Although a legal definition was put into federal law in 1990, sustainability is still more of a philosophy than a strict set of rules and as such is open to interpretation. But whatever the interpretation, a move toward sustainability supports seasonal, local foods.

I return to the March Farm booth for tomatoes and peaches. Emily and her husband, Tom Medonis, are now handling the table. In their twenties, they are blonde, tan, and fit. Emily carries their infant son, Cashlyn, in a sling as she waits on customers. She tells me she awoke at 4:30 a.m. to pick thirty pounds of blueberries.

Tom arranges peaches on the table. The fruit is pocked from a freak hailstorm in late June. I choose a half dozen and comment on the damage. Tom says, "The Europeans scoop these up. They don't care that they're damaged. This is what organic fruit looks like."

Damaged or not, the instinct to put a peach to her nose is too strong to resist for one customer.

"Is there any perfume better than this?" she says with a small sigh.

Another woman asks, ""Where do the peaches come from?" Tom answers without missing a beat, "From March Farm, ten minutes away from here."

The Litchfield Hills Farm-Fresh Market is run on all-volunteer basis. Kay has a list of sixty-five volunteers to call on, but relies on her three-person market master team plus a core group of six others, as well as "these wonderful teenagers who we couldn't run the market without." Kay's business experience includes new product development and strategic planning, and she ran her own business in transformational change before she "retired."

"We manage the market like it's a business. We know about advertising, we know about margins, we know about frequency, we know about research." Vendors have three different options for participation. They can sign up for the full nineteen-week season for $300, or a nine-week half season ($180 — they pick the weeks), or they can be a guest vendor for one to four weeks ($25 per week). "It doesn't make sense to us that we should be forcing people to take a full season," Kay says. "We'll end up with unhappy vendors and unhappy customers."

"The market aims to be a destination. "At Story Hour we'll have 25 children every week. The children learn, it draws people who stay around longer, and we grab the kids early. It's a family activity. Parents tell us that the kids wake up on Saturday morning and ask, 'When are we going to the market?' That's what it's about."

The market is so popular that customer demand generated an ongoing winter market, held in the Litchfield Community Center, a mile and a half down the street from the summer market. March Farm is there every Saturday. Music and agricultural-related activities are as much a part of the winter market as the summer. Because the community center has a good-sized kitchen, cooking demonstrations are more easily carried out, adding to the winter market's attraction.

My shopping is complete, yet I leave reluctantly. Although I have spent a little more than I planned, the produce is irresistible, and it feels right to support these hardworking producers. Shopping for food this way, I feel secretly European, sophisticated and earthy at the same time. I am tuned into the seasonal cycle, delighted by the pleasures of fresh food, anticipating delicious, healthy meals to come.

"Although this is only our fourth year, everyone says it feels like the market has always been around." Kay says, looking around at produce-laden tables. "They come for more than just shopping. They come for the experience."

I guess I'm not alone.

Websites

For a listing of Connecticut farmers' markets by county:
www.ct.gov/doag/

For information about using food stamps at farmers' markets:
www.cga.ct.gov/

Cornfield at Swendsen Farm in Bethlehem, where the Marches lease land to grow corn.

The Maze of Corn

The small corn maze sits at the top of the western orchards. On a gorgeous September day I drive the rough dirt road, following the new red-and-white signs directing me to the neatly mowed parking area marked by round hay bales. When I get out of the car, yellow green corn tassels are the only thing I see between me and the sky. Corn leaves rustle in the wind. It's a dry, eerie sound, reflecting the summer drought. Below me I hear the *b-r-r-r* of a tractor and voices of farmworkers in the orchard.

I walk into the maze. Posted at decision points are questions about the history and operation of the farm with multiple-choice answers. "How many kinds of apples are grown here at the farm?" Each answer sends you in a different direction, either straight, left, or right. The correct answer, C:14, takes you straight ahead to the next decision point. The only penalty for a wrong answer is that you either come to a dead end or circle around to where you began.

Corn mazes have become popular in Connecticut and nationwide. There are at least a dozen in this state and over six hundred in the United States. It's an innovative way for a farm to bring in additional income. Mazes can be elaborate, and GPS is sometimes employed to create theme designs, such as a huge Spider-Man or a larger-than-life John Wayne. Some feature detailed farm scenes or patriotic images. The March maze is simple, and it's free of charge to walk, an added activity for the growing agritourism element of the farm. "My goal is to make the farm a destination where people come and spend the entire afternoon," Ben March says.

Tom March drives by the greenhouses in his typical summer attire.

The corn grown for the maze is silage corn, known as cow corn. At the end of the fall, it will be cut down and chopped up to be used for livestock feed on area dairy farms. All the other corn grown at March Farm is sweet corn, for human consumption. Although it's the second biggest moneymaker for the Marches, most of the sweet corn is grown on leased farmland in town. One of these properties is the Swendsen Farm, a town-owned property on the east side of Bethlehem where Tom leases thirty-five acres.

A little past six a.m. on an August morning in 2005, Ben and three Jamaican farmhands enter a cornfield on the western side of the Swendsen Farm. Billy Collett, Ben's uncle, drives a tractor with an open wagon attached to the rear. The fresh-picked corn collected in the wagon will be on a dinner plate tonight.

A sea of tawny tassels over six feet high fills the view to the horizon. The early morning sunlight slants down through a break in the silvered sky, making a windowpane on the cloud cover. If you look down the open row in the middle of the field, the row left unplanted for tractor passage, the corn seems to march on forever. The corn silk spilling out of the ears is deep umber, an indicator of ripeness. The ears stand on the stalk like soldiers at attention. Thin veins stripe the snug green outer leaves.

The three farmhands wear yellow slickers, jackets, and pants. Although the air is still a little cool and the sky cloudy, they look hot. The slickers protect their skin, which is sensitive to the scratchiness of corn leaves and the dampness of the dew. Ben, unbothered, is in short sleeves and high boots and sweatpants cut off at the knees. Billy, in the tractor, wears a T-shirt, khaki shorts, and work boots. As the men move through the field, they look as if they're in the middle of a foreign jungle.

The men pick the corn by hand, then toss it out to the wagon. Billy paces the tractor to match their speed. Corn flies over the tops of the stalks like tiny fighter jets. They will work until the wagon is filled with enough corn to meet the day's orders, then sort and bag it on-site.

Like everything else on the farm, even with the use of machinery, picking corn by hand is labor-intensive. In 2006 the Marches decided to purchase a corn picker, an investment of $30,000. Ben's persuasive powers sealed the deal. "God forbid we would ever talk about buying a corn picker in 2004, when I first came back. I would have been laughed at. 'What do you think, we're millionaires here?'" he says. "But with a little research and planning, it wasn't out of the question, and we ended up making it work. We expect the picker to pay for itself after four to five years of use."

Ben's cousin, Shane, adds, "We had a corn picker fifteen or twenty years ago. My uncle (Tom March) picked once with it for about twenty minutes, then he parked it and started swearing and said the thing was a piece of crap. He's got no patience when things break down. You'll waste some corn with the picker, but if you're good with it, it will pick two ears per stalk." Waste is minimized because the efficiency of the picker allows corn to be picked in smaller quantities, as it ripens.

One summer morning, the new machine is in action in the small cornfield next to the upper blueberry orchards. The picker consists of a disc cutter set low to the ground connected to a tractor with hydraulic controls. Once cut, the ear is then carried up two angled rods to a conveyor belt, which delivers the corn to a flat open wagon being pulled alongside the picker by another tractor. Shane, driving the corn picker, watches the row he's cutting, head turning to the right then left. Ben, steering the tractor with the wagon, watches and matches Shane's pace.

The sun crests the corn tops. The light spreads wide across the field as the smell of diesel fuel mixes with the grassy smell of fresh-cut corn. Two quick passes and the bottom of the wagon is covered with corn. Pieces of cut cornstalk fly into the air. Once the machine stops, birds cover the cornfield to gather up leftovers.

"Have you seen the picker?" Tom asks me the next time I see him. "They picked 850 bags yesterday. That would take so much time last year." The highest volume of picking was 1,000 bags one September day, when Shane and Ben worked until ten in the evening to satisfy a last-minute order. All the picking is now done by the corn picker, unless fewer than 50 bags are needed for the day's orders, which is rare in corn season, when orders average between 300 and 500 bags daily.

Whether hand- or machine-picked, the corn is sorted and bagged by hand. Formerly done in the field, where there was no cover, the bagging operation is now situated in the shade of a plastic-covered greenhouse frame set up behind the store. Bagging is rough on the hands. Rocky wears gloves, but the others work bare-handed. Ben says he gets calluses for a couple of weeks, and sometimes his hands ache in the early September cold. The most intense period is the two to three weeks around Labor Day, when four men work all day every day bagging corn. Those days they order in breakfast, lunch, and dinner.

Five ears of corn are scooped up at time, and extra-long stems are snapped off. Sixty ears fill a bag. The work is done quickly, without talking, all business. The heat has not yet penetrated the plastic overhead, but it soon will. The Jamaicans wear baseball hats, dungarees, and long-sleeved shirts over their T-shirts. Reggae blasts from a boom box set up behind the farm store stand. They appear happy. I remember what Sue March told me about the workers. "They're willing to work more than forty hours. They're here to make money. They don't say, 'Oh, I have to go somewhere on a certain day.' They go home and use that money to run their own farms in Jamaica."

This bagged corn will end up in a number of local farm stands and at local wholesalers, where it will be delivered in the March Farm truck, a large box truck with the colorful March Farm logo painted on the side. Connecticut wholesale grocers include Jarjura and Sons in Waterbury, Bozzuto's in Cheshire, and Fowler &

Huntting (Fresh Point) in Hartford. In 2007 nearly three-quarters of a million ears of corn were sold wholesale. In addition, some corn will be sold in the March Farm store.

The corn season begins in late spring. Tom waits for the right conditions to harrow the soil, which needs to be a consistent fifty to fifty-five degrees to plant. In Connecticut, early spring is often marked by warm days but frosted nights. The soil must be plowable, so a rainy, wet spring can also delay planting.

If the elements have cooperated the previous fall, winter rye is planted at the Swendsen Farm and the Welicaitis Farm, where Tom leases an additional thirty-five acres for a total of seventy leased acres on other farms. By late March or early April the cornfields are golf course green with their winter rye cover, which is planted to help prevent soil erosion and to add nutrients to the soil when it is plowed under in the spring.

In October 2008, Ben decided to expand the corn maze into a haunted maze and opened it on weekend evenings for the public. On a chilly Saturday evening my husband and I enter the converted maze through a doorway erected from hay bales strung with tiny purple lights. We immediately faced an open grave, whose former occupant now roams the maze looking for those who disturb the corn. We are handed glow-light wands as we enter the darkness of the maze. I'm glad we have them.

Haunts and hollers and eerie sounds surround us. We hear rustling within the corn stalks but can't see anyone. When a white-faced apparition dressed all in black jumps out at a dark corner, I scream, clutching my husband's arm. The figures slip in and out of sight, like ominous shadows. The sound of a running chain saw startles another scream out of me, and then we are laughing. We make it to the end and buy a cup of warm cider at the concession stand, where we meet up with Ben and talk about his plans for next year, which include an expanded maze with more features.

On the walk back to our car, we stop to look over the farm under the full moon high in the sky. I think about Ben, full of plans, and his father, Tom, telling him when he returned to the farm, "Do whatever you want, but I'm not doing any more than I'm doing now." Tom, winding down as much as a farmer can, as if he's reaching the end of the farming labyrinth, has made all the turns and hit the dead ends and found the path that moves the farm forward. Ben, young and enthusiastic, embracing the farm, not knowing what's ahead of him but eager to find his own way through it.

Life as a farmer, which appears routine and predictable, is not unlike a maze. In the height of the season, the focus is on day-to-day operations and putting out fires, not the long-term perspective. At every decision point, which way? Right, left, or straight? For a farmer, a wrong answer can be costly. Money sunk into seed and fertilizer has to yield profits that year. Investments in equipment and orchard stock take longer to yield payback. Overhauling even a part of an operation can be risky and expensive. There are many unanswered questions. If only it were as simple as choosing A, B, or C, and circling back to where you began without concern for the consequences.

Grilled Corn

This is the easiest and tastiest way to prepare corn, and it couldn't be simpler.

- Husk corn.
- Place ears on hot gas grill; cover grill.
- Grill for 20 minutes, checking every 5 minutes.
- Turn ears so corn grills evenly.
- Serve — no butter needed.

Make sure to grill plenty of extra ears.

Corn Salsa

From Cynthia Rabinowitz

- 1 T extra virgin olive oil
- 1 T red wine vinegar
- Salt and freshly ground black pepper
- 3 cups fresh cooked corn
- 20 yellow or red cherry tomatoes or 1-2 cups chopped large tomatoes
- 1 T fresh flat leaf parsley, finely chopped
- 2 T fresh basil, finely chopped

Mix oil, vinegar, salt, and pepper, then combine with other ingredients. Let stand for 2-3 hours and serve at room temperature.

Spectrum of Agriculture

Modern agriculture ranges from "big ag" — think of the fully mechanized farms in *A Thousand Acres* by Jane Smiley — to the family garden. In between are varying grades of intensive and sustainable farming.

For third-generation farmer Tom March, farming the conventional way his father and grandfather did has been the accepted practice. The use of chemical pesticides and herbicides was part of modern agriculture. Fourth-generation farmer Ben March has never liked the chemicals and is evaluating alternatives. His hope is to be completely organic in five years.

"In ten to fifteen years, they may discover that the chemicals we use now are bad for you," he says. "I feel directly responsible for the farm now." He is beginning his assessment of what March Farms can do differently starting with one of its largest crops, greenhouse tomatoes.

The number one cause of tomato disease is humidity, he tells me. Tomato plants perspire to cool off, like humans do. Because the leaf surface remains ten to twenty degrees warmer than the air, the plants continue to sweat for two to three hours after dark. Moisture left on the leaf surface causes botrytis and powdery mildew.

Ben worked with the University of Connecticut Cooperative Extension System to establish a humidity control system. By monitoring humidity and temperature and venting the greenhouses as needed, the humidity can be kept close to 80 percent, which is "golden," Ben says.

At the end of each day, Ben vents the farm's greenhouses until 65 percent humidity is reached, then the exhaust fans at each end of the greenhouses are shut down. The cold air is reheated by the outdoor wood stoves stationed between greenhouses, which makes the air absorb the humidity, until the humidity is steady. In early March, Ben does laps between the greenhouses from 5:30 to 6:00 p.m., turning on fans, checking humidity and temperature readings. Once the environment is stabilized, he returns at eight to reload the outdoor wood stoves and vent again. His goal is a temperature range of 62 to 65 degrees, with 80 percent humidity. Any higher temperature wastes the photosynthetic processes.

To automate this process, Ben's brother, Tom, is installing a relay system. Once active, the relay system will shut off the stove once the humidity hits 75 percent. A five-minute delay allows the stove to exhaust fumes. Exhaust fans kick on until the humidity and temperature are in the desirable range, then shut off. Eventually all the greenhouses will be automated, and Ben will have to find another way to get his evening exercise. This system may burn a bit more oil and wood, but it has eliminated the need for chemicals. So far this year, the Marches haven't had to spray once. About five years ago they converted to a biopesticide, which is derived from natural materials such as animals, plants, bacteria, and minerals, to deal with the fungi. The organic fungicide they've used is Actinovate, a beneficial bacterium.

The plants are also benefiting from a new drip irrigation system. Although it now takes longer to water, they use less water and, even better, no water goes on the tomato leaves.

"I've never seen the plants look healthier, and there's been no petal drop," Ben says.

March Farm sign, on the corner of Route 61 and Bellamy Lane, just north of the center of Bethlehem.

"Last year, which was so wet, I took wheelbarrows of tomatoes dropped from the plants from stem rot. Now we have a higher yield and a healthier environment."

I ask how his father, Tom March, views what he's doing.

"My father wanted to add another greenhouse this year, but I convinced him to try this," Ben says, "see if we can optimize our current greenhouses before adding another one."

"Is he convinced?" I ask.

"He checks with me every night." Ben smiles. "He always wants to know the readings in the greenhouse. I guess that means yes."

The techniques Ben has utilized in the greenhouse are part of integrated pest management (IPM), which employs, according to the USDA website, "diverse methods of pest controls, paired with monitoring to reduce unnecessary pesticide applications." A combination of approaches is used to manage pests with minimum negative effects.

In Connecticut IPM is supported by UConn's College of Agriculture and Natural Resources, which provides technical assistance and training to growers, groundskeepers, homeowners, and students. The program's mission is twofold: supporting the farmer and protecting the environment. Methods include a variety of controls, such as beneficial organisms, traps, crop rotation, resistant crop varieties, perimeter trap cropping, use of chemicals, and physical barriers.

Implementing IPM takes extra time. The farmer has to keep records and track crops. Tasks involve setting traps, keeping records, using the CTDOA's early warning service, and spraying, but only when necessary. Ben and Tom have time to do this now because the newer generation of Jamaican seasonal workers, hired from April through October, want more responsibility, freeing up Tom and Ben's time to monitor pest activity on the farm and then follow up quickly.

It's been relatively easy to eliminate the chemicals used on tomatoes. Corn is another story. Known as a "heavy user," corn not only draws nutrients from the soil but, because of its vulnerability to many insects, requires application of numerous chemicals. These pesticides may leach from the soil and contaminate the groundwater. At least one IPM practice is in place with corn already: waiting to spray the corn until visible evidence of earworms.

Two different kinds of chemicals are used on corn: herbicides and pesticides. Herbicides are applied once a year, about four days after planting, to prevent weed growth, mostly crabgrass and quack grass. Pesticides battle a number of insects, including earworms, corn borers, and armyworms. In late spring, Tom sets moth traps for earworms. Once he sees moths in the traps, he'll start spraying. He will also call the IPM number to find out what CTDOA staffers have seen in the Connecticut River Valley farms in the center of the state, where the moths often show up a few weeks before they do in the northwest. He then begins his spraying routine, once or twice a week until September, when the corn silk turns brown. The chemical he uses, Warrior, will prevent not only earworms but also corn borers and armyworms. In addition, Tom takes a soil sample annually to determine if he'll need to put down lime in the fall and/or fertilizer the next spring. If lime is used, the average application is one to two tons per acre.

Ben March adding fertilizer to the water tank that feeds the greenhouses.

Sweet corn is always in demand by consumers, but it's becoming harder for farmers to find landlords. Some local land trusts have banned farmers from leasing land for corn crops because of the need for intensive application of chemicals. Hay is much more popular with land trusts, as it's less demanding of soil nutrients and chemicals. Tom leases three properties near his own farm. The Bethlehem Land Trust allows him to grow both hay and sweet corn on its Bellamy property, the town of Bethlehem leases him land on the Swendsen Farm Preserve to grow both hay and corn, and he leases acreage at the Welicaitis farm to grow corn.

The fruit orchards have always been sprayed with pesticides to contain damage done by insects and with herbicides to control fungi. When Tom's father, Matt, mixed vats of chemicals for spraying, he'd use his bare arm to stir the liquid blend, or so the story goes. Back in the forties little was known about the effects of pesticides. In a scene from *A Bountiful Century: Fruit Growing in Connecticut, 1891–1991* (University of Connecticut), a documentary detailing the history of Connecticut orchards, farmers are shown in short sleeves and straw hats driving open tractors without protective cabs as they sprayed heavy clouds of chemicals on their fruit trees. In 2000 Tom began using a "smart" sprayer, which is computerized to gauge the dimensions of the tree and allot the appropriate amount of chemical to be delivered. He's driven an enclosed cab tractor since 1987.

Spraying begins in the spring when the blossoms are out and continues about every two weeks until three weeks before picking. A fungicide is applied to prevent scabs from forming in the leaves, which can spread quickly once established. The next application is an insecticide for the apples to make sure that plant bugs, also known as mirids, don't ruin the apples by biting them at the blossoms. The bug is prevalent along hedgerows and near woodlands, a common landscape on the farm. At the same time, the peaches become vulnerable to brown rot and require a fungicide. Tom uses Pristine, but you have to stay out of the orchard for at least twelve hours after it's applied.

"Even the organic sprays require twelve hours before re-entry," Tom says.

Mites are another problem in the orchards, but they primarily affect only the leaves of the trees, not the fruit. Tom watches if the leaves turn bronze, then he applies miticide only once a year, as it's expensive. Once he sees it on one section of the orchard, the entire orchard must be sprayed because mites spread so quickly.

I catch up with Ben in early October, as he is putting the final touches on the haunted graveyard, a Halloween attraction for younger children, installed where a sunflower maze stood earlier in the season. I ask him about the tomato crop.

"Getting the humidity under control increased the yield tremendously. It was the equivalent of having four additional greenhouses. I hoped to have time to assess the corn and apples this year, but I was so busy keeping up with the tomatoes I didn't get to it. As the greenhouses become automated—half are there already—I'll be able to address the corn and apples. Next year." He smiles as he says it. It's clear he understands the working rhythm of a farm, how priorities are constantly shifting.

"And the strawberries aren't sprayed at all, except for a foliar fertilizer feed," he says. "We don't spray the cherries either. For the blueberries, we only use Roundup once, early in the season.

"I'm excited about the apples. We bought a new mower with a hydraulic arm that wraps around the trunk of each tree and mows completely around the base of the tree. Now we don't have to apply Roundup in the orchards. One less chemical. We're getting there."

CHAPTER FIVE

Looking Forward

The Future of the Farm

For decades payment at the March Farm store was based on an honor system. The store was open all the time. When a customer came in and no one was available to ring her up, she chose her produce, weighed it on a simple digital scale, and made change in the green plastic basket that sat on the counter. Customers sometimes wrote down the cost of their purchase on a scrap of paper, but often not. Recently that system has been eliminated because people have been stealing from the farm.

From Memorial Day through October the store is manned seven days a week. In the height of the pick-your-own season, customers used to come into the store, grab a basket, and head out to an orchard. They would return, have their produce weighed, and pay. Now there is different procedure: when you pick up your basket, you exchange your car keys or a twenty dollar deposit for a numbered ticket before you're handed a basket and go to the orchard.

"People would leave with six baskets and come back with four," Tom explains. "Now, it's 'Gotta turn in your ticket.' I hate that it's this way."

This isn't the only thieving that's been going on. At least once, someone driving a vehicle with New York plates drove up to the orchard and picked bushels of apples then took them to a farm stand and sold them.

As March Farm's popularity reaches a wider range, people are coming from farther away to visit. At five o'clock on a Columbus Day weekend, the parking lot above the store is jammed. Children on pedal tractors fill the hay bale playscape; their parents watch from nearby picnic tables in the shade of the newly built pavilion. People are crawling over the orchards gathering apples. Mum plants displayed outside the store are moving fast; children are picking out pumpkins laid out on hay next to the store. Across the street, the goats and sheep next to the pond are getting plenty of attention. They bray and kick within their pen. Walkers head around the pond and over the wooden bridge to hike up into the orchards and take in the view.

This is agritourism, giving people an experience of being on the farm. Many farmers are developing this aspect of their business in order to bring more people to the farm, people who are hungry not just for the produce, but

Above: The playscape at March Farm, a popular family spot. Opposite: Feeding the animals in the petting zoo across the street from the playscape.

Left: Picking blueberries in the lower orchard. Right: Goats in the petting zoo share their perch.

for the farm environment. It also provides an inexpensive family activity. The volleyball court area, complete with grills and paddleboats, is now available for party rentals. This area is also the site of family parties on holiday weekends, as it has been for decades.

The Marches are beginning to see the correlation between agritourism and increased profits. In 2009, when the wet summer halved the income from corn, usually one of the bigger crops, and the apple crop was shot by an intense June hailstorm, the retail store never did better. The website (www.marchfarms.com) recorded 63,000 unique individual hits in 2009. This number does not include repeated visits. Although people arrived at the site by searching for a pick-your-own farm, when they got there they first clicked on the playscape page, then the Farm Life Fun page, and finally went to the pick-your-own page.

Plans for expansion of agritourism continue. In 2010 the playscape was expanded to four times its size and, close by, a sunflower maze was also added. An architect is drawing up plans to expand the store.

Late in 2009 Tom and Sue applied to the state of Connecticut for the purchase of development rights on their farm. In 2010 the two parties came to an agreement: the state will purchase the development value of the farm, giving the Marches a financial settlement in exchange for an agreement to keep the farm in agriculture from this time forward. Tom and Sue will still own the farm, but they are giving up the right to sell it to a developer, most likely at a higher price. The sale of development rights carries with it a deed restriction that guarantees the farm will stay in agriculture for perpetuity. The state provides fifty percent of the funding; a matching grant program funded through the Natural Resources Conservation Service, an arm of the USDA, provides the other half.

Tom tells me he's spoken to all his children and they're fine with the arrangement. Acreage around the house and store will be kept separately in the Marches' name.

When the people from the state came out to talk to the Marches, they said it's a "perfect farm" for the program.

"Two days later we got a card from the woman," Tom says.

"Isn't it nice to be told your farm is perfect?"

Tom grins and wipes his hand across his face. "Yeah."

"I'm sixty-five," he tells me. "In 2015 the farm will be one hundred years old."

"Are you collecting Social Security?" I ask.

His smile is wry. "I get $1,300 a month, and I hand it over to Sue, who gives me $300 back for pocket money. The rest goes into the farm."

Although the farm's popularity has exploded over the past few years, neighbors and townspeople have long had an emotional stake in seeing it succeed.

Jennifer Hunt has lived close-by for over twenty years. Her sons, Alex and Simon, grew up as playmates of the March children.

"It's almost like I'm a part of it. Whenever we travel, we always bring something from March Farm as a gift," she says. "I've watched Tom take over from his father and the farm expand over the years. It just warms my heart to see Ben come back and know the farm is secure for the next generation."

Even those without personal ties feel connected to the farm. A young woman from nearby Southbury tells me on a farm hike, "This is my favorite place in Connecticut."

And Tom tells me he overheard two women talking in the farm stand one day, one a Bethlehem resident, the other from a nearby town. The Bethlehem resident said to her friend, "We have something you don't have. We have March Farm."

Enough For One Life

When snow covers the landscape, the farm buildings at March Farm stand out like pencil sketches on bleached paper. The farm store is quiet, empty of all but the occasional customer. The apple house, too, is empty; a few boxes of apples remain. Machine parts are strewn on top of the conveyor belt where the autumn apples are washed, polished, and sorted. All is utility here on the farm, no sentimentality. Even in the winter, there is no time for it.

But the greenhouses are pregnant with new growth. Their plastic glows in the weak winter sun. Inside is the seed of the next crop of tomatoes, tiny three-inch plants that in a few weeks will be a half foot high.

Farmer Tom March has a large Band-Aid across his cheek as he checks the seedlings. He speaks of a skin cancer on his face, removed successfully, almost as if he's amused by such a thing. Listening to him I wonder, has he ever been seriously sick? Or is illness a luxury to farmers?

I remember how Tom looked last summer, hopping out of his truck at Swendsen Farm here in Bethlehem, where he leases acreage for corn. He wanted to move a boulder off the farm road. As he bent over, his ropy thigh muscles were exposed from under his cutoffs, and it struck me how incredibly strong he is. He is always on the move, even eats lunch standing up at the kitchen island. His stamina appears unending.

In winter the pace at most Connecticut farms eases a bit, but in the orchards there is still pruning to be done, each tree tended to individually, by hand. There is always greenhouse activity, a cycle renewing. Farming is tied to such rhythms. They preserve a way of life, a life not just about food production, but about the essentials. A farmer knows simple rewards: the sweet satisfaction of applying physical and mental efforts to an endeavor bigger than oneself.

This is the farm where Tom grew up, where his father worked the land, the farm his grandfather bought in 1915, after arriving from Lithuania. Tom left the farm only once after high school, to attend the University of Connecticut. He's never taken a vacation, or had any desire to. He and his wife, Sue, are now looking toward the future, a hundred-year anniversary in 2015, the passing of the farm to the fourth generation. Their son Ben and nephew, Shane, now work full-time at the farm, ensuring the family farm will continue. The recent sale of development right guarantees the land will remain in agriculture.

What I am most struck by here at March Farm is the sense of stewardship, of quality, of integrity. These are people who know who they are, know their place in the world and the work they have to do. They go about it in a straightforward and committed way. This seems, to me, enough for any one life.

∾

Opposite: Farm pond, fed by Long Meadow Lake, provides water for the strawberries and blueberries grown across the street, as well as being a source of recreation.

ACKNOWLEDGMENTS

With gratitude to:

The March and Collett families and all those who agreed to be interviewed for the book; Allison Davis, whose enthusiasm never flagged; Mary Devine, who held the vision of the book; Sue Meister, whose writing support came at a critical time; Mary Moore, who read every word and whose suggestions improved the book immeasurably; Kimberley Reynolds, whose friendship and support sustained me; Jean Sands, who started me writing; Linnie York, who never stopped believing in me and the book; Jack Huber, book designer angel, who made the book more beautiful than I ever imagined it could be; Melissa DeMeo, copy editor; Sian Hunter, editor; Kathleen Spivack, poet, writer, teacher; Shepaug River Writers Group, for camaraderie, support, and many happy memories; and friends and family who encouraged and supported me through the journey.

Special thanks to Sisters in Script for funding. Sisters in Script (www.sistersinscript.org) offers an annual grant to a woman writer who is self-publishing her first book. I am grateful to be the 2011 recipient.

And most importantly, to Doug, without whom none of it would be possible.

A version of "Enough for One Life" appeared in the winter 2010 issue of *Edible Nutmeg*.

A version of "Blueberries" appeared in the August 2003 edition of the *Litchfield County Times*.

Opposite: Apple orchard, mid-summer.

SOURCES

Introduction: The Present

Stephen R. Kellert and Edward O. Wilson, eds., *The Biophilia Hypothesis* (Washington, D.C.: Island Press, 1993).

Theodore Roszak, Mary E. Gomes, and Allen D. Kanner, eds., *Ecopsychology: Restoring the Earth, Healing the Mind* (San Francisco: Sierra Club Books, 1995).

Farmland Preservation Program, Connecticut Department of Agriculture, www.ct.gov/doag.

Michael Pollan, "Weed It and Reap," *The New York Times,* November 4, 2007.

The Art of the Commonplace: The Agrarian Essays of Wendell Berry, ed. Norman Wirzba (Berkeley, Calif.: Counterpoint Press, 2002).

Plow to Plate, New Milford Hospital, www.plowtoplate.org.

Yale Sustainable Food Project, Yale University, www.yale.edu/sustainablefood.

Bill McKibben, Deep Economy: *The Wealth of Communities and the Durable Future* (New York: Times Books, Henry Holt and Company, 2007).

Michael Pollan, *In Defense of Food: An Eater's Manifesto* (New York: The Penguin Press, 2008).

Rick Levine et al., *The Cluetrain Manifesto: The End of Business as Usual,* 10th anniversary edition (New York: Basic Books, 2011).

Chapter One

In the Apple House

Ronald Jager, *The Fate of Family Farming: Variations on an American Idea* (Lebanon, N.H.: University Press of New England, 2004)

Farm Field Trip

Paul Singley, "Learning, Sweet as Apple Cider," (Waterbury, Conn.) *Republican-American,* October 26, 2005.

Kellert and Wilson, *The Biophilia Hypothesis.*

Richard Louv, *Last Child in the Woods: Saving Our Children from Nature-Deficit Disorder* (Chapel Hill, N.C.: Algonquin Books, 2005).

No Farms, No Food

"Oxford Word of the Year: Locavore" (blog), Oxford University Press website (blog.oup.com/2007/11/locavore), November 12, 2007.

Kim O'Donnel, "A Mighty Appetite" (blog), *Washington Post,* September 19, 2006.

Park Slope Food Coop, www.foodcoop.com.

Tom Philpott, "Latest E. coli outbreak should prompt rethink of industrial agriculture," Grist, September 21, 2006, grist.org/article/e-coli.

Earthbound Farm, www.ebfarm.com.

Philip H. Howard, www.msu.edu/~howardp.

Edible Nutmeg, www.ediblecommunities.com/nutmeg.

Connecticut Grown Program, Connecticut Department of Agriculture, www.ct.gov/doag.
buyCTgrown, CitySeed, buyctgrown.com.

Samuel Fromartz, *Organic, Inc.: Natural Foods and How They Grew* (City: Houghton Mifflin Harcourt, 2006).

Jennifer Maiser, "10 Reasons to Eat Local Food," Eat Local Challenge (group blog), www.eatlocalchallenge.com.

Above: Farmhand Ken Sperry, early 1970s. Opposite: Nathan Bloss Farm around 1890, bought in 1915 by Thomas Marchukaitis.

Top: Matt March in the apple orchard, 1980s.
Above: Tom's mother, Anastasia, 1935.

Chapter Two

History of March Farm

Michael Bell, *The Face of Connecticut: People, Geology, and the Land,* State Geological and Natural History Survey of Connecticut (Hartford, Conn.: Connecticut Department of Environmental Protection, 1985).

Definition of prime farmland soil, Natural Resources Conservation Service, www.ct.nrcs.usda.gov.

Farmland Information Center, www.farmlandinfo.org/agricultural_statistics.

Working Lands Alliance, American Farmland Trust, www.workinglandsalliance.org.

Old Bethlem Historical Society, Bethlehem, *Connecticut: A Primer of Local History from the Beginning to 1876* (Bethlehem, Conn.: Old Bethlem Historical Society, Inc., 1976).

David M. Melesky, *I Remember Bethlehem* (Greenville, Pa.: Birch Tree Ranch, 2008).

Marshall Linden and Linton E. Simerl, eds., *250 Years of The First Church of Bethlehem* (Bethlehem, Conn.: First Church of Bethlehem, 1990).

Carol Ann Brown, *Bethlehem,* Images of America series (Mount Pleasant, S.C.: Arcadia Publishing, 2009).

Winter Pruning

"Training & Pruning Fruit Trees," North Carolina Cooperative Extension Service, www.ces.ncsu.edu.

George Kuepper, Guy K. Ames, and Ann Baier, "Tree Fruits: Organic Production Overview" (Butte, Mont.: National Center for Appropriate Technology, 2004).

Blizzard of 1978, Bethlehem Style

Daniela Altimari, "Feb. 6-7, 1978: The Blizzard of '78 Shut Down the State and Made Heroes Out of Those with Four-Wheel Drive," *Hartford Courant,* February 25, 1998.

Tom March, "Blizzard of 1978, Bethlehem Style" (Bethlehem, Conn., 1978).

Chapter Three

Jamaican Farmhands

"Jamaica," Infoplease, www.infoplease.com/country/profiles/jamaica.html.

Joseph Young, executive director of the New England Apple Council, interview by the author.

Julia Preston, "White House Moves to Ease Guest Worker Program," New York Times, February 7, 2008.

William Kandel, *Profile of Hired Farmworkers, A 2008 Update* (Washington, D.C.: Economic Research Service, 2008).

Eduardo González, Jr., "Migrant Farm Workers: Our Nation's Invisible Population," eXtension, May 27, 2008, www.extension.org/pages/9960.

Tom Philpott, "How globalization is smothering U.S. fruit and vegetable farms," Grist, August 30, 2007, grist.org/article/worldfood.

We Were Due

David Mas Masumoto, *Epitaph for a Peach: Four Seasons on My Family Farm* (San Francisco: HarperSanFrancisco, 1995).

Keeping the Farm Afloat

Sue March, interview by the author, March 17, 2008.

Chapter Four

Farmers Market

Litchfield Hills Food Systems, Inc., litchfieldhillsfarmfresh-ct.org.

Kay Carroll, market master of Litchfield Hills Farm-Fresh Market, interview by the author, August 5, 2009.

"Connecticut Farmers' Markets," Connecticut Department of Agriculture, www.ct.gov/doag.

"Farmers Markets and Local Food Marketing," Agricultural Marketing Service, U.S. Department of Agriculture, www.ams.usda.gov.

Edward Ragland and Debra Tropp, "USDA National Farmers Market Manager Survey 2006" (Washington, D.C.: U.S. Department of Agriculture, Agricultural Marketing Service, May 2009).

Sustainable Table, www.sustainabletable.org.

Spectrum of Agriculture

Jane Smiley, *A Thousand Acres* (New York: Alfred A. Knopf, 1991).

"Integrated Pest Management," National Institute of Food and Agriculture, U.S. Department of Agriculture, www.csrees.usda.gov/integratedpestmanagement.cfm.

Integrated pest management, Connecticut Cooperative Extension System, College of Agriculture and Natural Resources, University of Connecticut, www.extension.uconn.edu.

Lucy McTeer Brusic and Tom Capistrant, *A Bountiful Century: Fruit Growing in Connecticut, 1891–1991* (New Haven, Conn.: Connecticut Pomological Society, 1990), VHS cassette.

Chapter Five

The Future of Farm

Natural Resources Conservation Service, U.S. Department of Agriculture, www.nrcs.usda.gov.

Above: Thomas and Rose Marchukaitis in their later years. Below: Matthew and Anastasia (Skeltis) March, 1955. Left: Original farmhouse on Nathan Bloss farm. It remained the main farmhouse on March Farm until 1974, when it was replaced with the current house, where Tom and Sue reside.

Above: Tomato seedlings in the greenhouse. Right: Tom March points out where the root stock, which determines the final size of the tree, is grafted to the cultivar above it. Opposite: View from apple orchards along Munger Lane. Next page: Corn is picked daily to be sold fresh in the farm store and through wholesale markets.